DARK SKIES

A Journey Into The Wild Night

Tiffany Francis

BLOOMSBURY WILDLIFE
LONDON • OXFORD • NEW YORK • NEW DELHI • SYDNEY

To my sister and number-one fan, Chloë

BLOOMSBURY WILDLIFE
Bloomsbury Publishing Plc
50 Bedford Square, London, WC1B 3DP, UK

BLOOMSBURY, BLOOMSBURY WILDLIFE and the Diana logo are trademarks of Bloomsbury Publishing Plc

First published in Great Britain 2019

A catalogue record for this book is available from the British Library

Library of Congress Cataloguing-in-Publication data has been applied for.

ISBN: HB: 978-1-4729-6459-5;
ePub: 978-1-4729-6457-1; ePDF: 978-1-4729-6458-8

2 4 6 8 10 9 7 5 3 1

Cover illustration by Tiffany Francis

Typeset in Bembo Std by Deanta Global Publishing Services, Chennai, India
Printed and bound in Great Britain by CPI Group (UK) Ltd, Croydon CR0 4YY

FSC www.fsc.org MIX Paper from responsible sources FSC® C020471

To find out more about our authors and books visit www.bloomsbury.com and sign up for our newsletters

Contents

The stars are forth, the moon above the tops
Of the snow-shining mountains. Beautiful!
I linger yet with Nature, for the night
Hath been to me a more familiar face
Than that of man; and in her starry shade
Of dim and solitary loveliness,
I learn'd the language of another world.

Manfred, George Gordon Lord Byron

TIF

CHAPTER ONE

Witching Hour

One night in late September I happened to be on the north Norfolk coast, cocooned in duvets in the back of my car, like a lost human burrito. I'd driven to Snettisham, an RSPB reserve formed of lagoons and mudflats that sprawl out to the Wash estuary and into the black North Sea. Here the stars were reflected in the water like pearlescent shoals, and the next morning I would watch waders sweep across the lagoons in the sunlight, my stomach full of blackberries foraged from the reserve paths. But tonight there was only a curlew bubbling, the sour scent of kelp and mud drifting in through the open window, and somewhere an oystercatcher *peep-peeping*; with those long, orange beaks they look like Pingu mid-*noot*.

Two days ago my relationship had ended. After all the tea was poured, and the conversation came to its natural end, I packed a bag and drove to East Anglia. The next morning my boyfriend and his brother would be going to France on the trip we had organised together, and the thought of lingering on in Hampshire was enough to send me instead to Norfolk, a temporary distraction from the loneliness that had started to creep into my body. I kept driving until I reached the sea, found some company in a swarm of wading birds, and spent the evening watching shelducks sift for invertebrates on a shingle beach.

By midnight a waning moon hung in the sky, and I lay awake listening to waves crash down onto silted sands. Outside the birds slept and the earth continued to revolve, so indifferent to the individual creatures in its web: the ocean was alive, the stars exploding, the soil murmuring with blood flow. The earth doesn't care for relationships and heartache, only the rotation of sun and moon, the eternal movement of life and living things. Everything we do depends on the sun rising every day, but half of our lives are spent in darkness. How much energy continues to burst from the landscape after the sun goes down? And by giving in to sleep when the world grows dark, how much of life are we missing out on?

I'm someone who crawls into bed each night and falls asleep within seconds. No restless hours of tedium turning over the day's thoughts like a tumble dryer; I surrender wholly to sleep, and my mind falls into that euphoric chasm of the unconscious.

Or so I thought.

Like many children, my parents divorced when I was in primary school, and I gained a new stepsister, Christie, who I now love like a blood sister. We helped each other through those limbo years between infancy and adolescence when you're unsure if you want to climb trees or message boys online. The best part of gaining a new sister was sharing a room at weekends, and it was at this point in my life, having never shared a bedroom before, that I discovered I am a somniloquist: I talk in my sleep.

It was and still is, hilarious. Conversations ranged far and wide; sometimes I would shake Christie awake to tell her 'zombies were coming', and more than once I talked about cake. I would laugh and shout and jump up in the middle of the night, convinced that spiders were crawling around my bed. It hasn't gone away with age. New sleep mates are forewarned. At university, I once crept over to my money box on the shelf, took it down and placed it on the floor. On returning to the bed and my boyfriend's puzzled expression, I replied: 'Burglars.'

I rarely remembered anything I had said or done in the night, unless it was so dramatic that I woke myself up in the middle of it, like the time I pulled the duvet off the bed and ran into the bathroom, only to wake up where I stood, bleary-eyed and confused. When I was first told what I had been doing, it came as a genuine surprise. But it also explained one or two mysterious occurrences I remembered from my childhood, like times I had woken up in the morning at the opposite end of the bed, or when the giant teddy that lived in my toy box had inexplicably ended up squashed against my head in bed. I can't explain the feeling of shifting from the unconscious mind to the conscious one, but it's incredibly disorientating and unpleasant, followed by an awkward conversation with your sleep mate who has been rudely awoken by a madwoman raving about spiders or loading the dishwasher.

After years of this behaviour I decided to see if the doctor could shed any light on my condition, and in 2015 I was referred to Papworth Respiratory Support and Sleep Centre in Cambridge, where I stayed the night under close observation. The other patients suffered from insomnia and narcolepsy, sharing weary tales of a life spent in drowsy fatigue. At night we were plugged into a machine and recorded on camera while we slept, and the next day I was eating toast and jam when my doctor arrived to ask if I thought I'd been 'active' in the night. I replied, over-confidently, that I had not, and he showed me a movement chart that looked like Jackson Pollock's latest masterpiece. I left the centre that afternoon with the advice to 'get more sleep' and decided to accept my flaws.

'What hath night to do with sleep?' asked John Milton in his poem *Comus*, named after the Greek god of nocturnal frivolity. Since the earliest days of human civilisation, we have evolved to live our lives in daylight and hide away at night, and there are logical reasons for this: our ancestors were preoccupied with survival – hunting and being hunted, gathering plants, building, socialising, farming and sleeping, all of which were entwined with the rhythms of day and night. But is it realistic to assume that of all the animals, we alone chose to sleep from dusk to dawn in undisturbed blocks? To suggest that older societies slept away the darkness and never enjoyed the midnight hour?

In the early 1990s, Dr Thomas Wehr carried out an experiment in human photoperiodicity – how the body reacts to day and night – at the National Institute of Mental Health in Bethesda, Maryland. His team were trying to recreate prehistoric sleeping conditions in a group of young men from Washington by controlling the hours they were exposed to daylight. While the modern American's typical day measures 16 hours, the subjects' days were shortened to just 10 hours, a more realistic idea of winter daylight for people without artificial light and coffee. For one month, the men came to the laboratory each night to spend the remaining 14 hours in dark, windowless rooms, encouraged to relax and sleep as much as possible. In measuring their brainwaves, temperature and hormone levels, the researchers recorded a change in their nocturnal behaviour.

As the subjects adjusted to their new schedule, their sleeping patterns shifted into two distinct phases. While they only slept for an hour more than usual, the total sleep time was spread over a 12-hour period and split into two sessions. The 'first sleep' was a deep, slow-wave sleep, lasting around four to five hours before they awoke around midnight for an interval of quiet reflection and relaxed wakefulness, described by the researchers as a state of meditation. The subjects then returned to a 'second sleep', characterised by rapid eye movement and vivid dreams, before waking naturally after four to five hours. Dr Wehr guessed that this rhythm of sleep was much closer to that of our ancestors who, like many other mammals, slept and woke in phases while never leaving the safety of their dens. During the meditative waking stage, the researchers also recorded an increase in prolactin, a compound that encourages an animal to rest; high levels are usually found in nursing mothers and chickens brooding their eggs for long periods of time. The men also released increased levels of the growth hormones that help the body to repair itself, which reduced dramatically when the subjects returned to their regular schedules.

If this is how we are inclined to sleep, it didn't disappear with our ancestors. Written records show that the 'first' and 'second' sleep was still practised far into the early modern era, like in this piece by Stuart poet George Wither:

> Think upon't at night,
> Soon in thy bed, when earth's depriv'd of light:
> I say at midnight, when thou wak'st from sleep,
> And lonely darkness doth in silence keep
> The grim-fac'd night …

The philosopher John Locke wrote that 'all men sleep by intervals', as did the Catalan philosopher Ramon Llull when he mentioned the *primo sonno* (first sleep) stretching from mid-evening to early morning. Memoirs, journals, letters, poetry and medical studies all refer to the idea as if it were common knowledge, and list a range of activities carried out at midnight. Men would smoke, write, eat and drink, while women completed light chores like carding wool, washing clothes, processing flax and, according to one maidservant, 'brewing a batch of malt'. Sexual encounters were inevitable, as well as poaching, stealing and witchcraft; one Italian charcoal burner in Ferrara witnessed his wife leave the bed at midnight and anoint herself from 'a hidden vase'.

The witching hour, the graveyard shift; why does the night remain so eerie to us? From an early age, we are told ghost stories set in dark places, warning us not to stray into the night. It feels frightening, macabre, even criminal, but the darkness also evokes a peace and solidarity with the landscape that fades away in sunlight. Perhaps there is something within this hidden world that can reconnect us with our landscape, at a time when we are neglecting our natural roots in favour of modernity's unnatural rhythms.

After Dave and I broke up, I moved into my sister's house with my sister, brother-in-law, two-year-old niece Meredith, and Bryan the cat. It was warm and cosy, but a two-year-old makes it difficult to concentrate on anything before she goes to sleep at 7pm, and I spent the autumn nights awake to catch up on writing, drawing and anything else that had been superseded by Meredith's new pirate ship or 70 episodes of *Hey Duggee*. It was as close to a 'second sleep' as I could get without physically going to bed at 7pm. The evenings were relaxed and peaceful; I ate cheese on toast and took long, hot showers.

As the weeks passed and I became used to sleeping alone, not only did I spend more time working at night, I noticed that when I chose to go to bed early, I started experiencing my own 'first' and 'second' sleep. It happened during the Christmas holiday when I had two weeks to lounge about with no schedule, and I wasn't prohibited by a nine-hour workday or attempts to squeeze freelance work into my spare time. Slow days spent quaffing gin and munching on strawberry cremes would float by, and the laziness of the season would call me up to the unconditional love of my duvet around 9pm. Asleep within seconds, I would later wake up in the early hours for no reason at all. No child crying, no storms, no hungover organs gasping for water. My mind just woke up.

At first, I did what most under-forties would do, and browsed social media until I fell back to sleep. I thought it was a one-off; probably underlying anxiety or a noise in the house that had dragged me from the depths of slumber. But then it happened the next night and the one after that. By the fourth, I realised social media is a depressing place to be in the early hours when your friends are asleep, and Donald Trump wakes up, and by the fifth, I decided it was time to leave the bed. The window of the spare room looks down onto the garden, a small but healthy patch with a plum tree, a vegetable plot and plenty of birds. That winter brought different species to the garden in search of nourishment, and I loved watching them from the warmth of the kitchen. There were gangs of long-tailed tits dangled from the plum tree like lollipops and a pair of blackcaps that might have migrated over from Germany to overwinter in Hampshire. I followed them around the garden with my camera, desperate to capture their black and chestnut heads shining in the December sun.

One morning in the early hours, before dawn arrived and the blackcaps and goldcrests were still silent in their nests, I walked downstairs and into the kitchen to find my wellies, opened the back door and stepped out into the cold night air.

At this hour, the lack of background noise from the human world means the air is of a different quality, which is why birdsong can carry 20 times further at the start of the dawn chorus. I stood in the garden with the cold pushing down on my warm skin, filling my lungs with the bitter air that I would have to share with the snoozing plum tree tonight if I wanted to continue breathing tomorrow. In the bordering shrubs, hedgehogs were hibernating. I couldn't hear them, but we watched them grow through the summer and saw them take shelter in the heap of garden waste by the shed. On spring nights they would emerge and gorge on the syrupy slugs that orbit the vegetable patch, but for now, they were warm and compact, a clutch of prickled eggs lying dormant beneath the decaying leaves of autumn.

The world was asleep, and above it all, the stars shone like broken glass.

Society today is shaped by eight-hour working days, but we are continually being told to spend more time outdoors to improve our physical and mental health. Should we be making use of our nocturnal freedom to explore the landscape at night in peace, to see everything from a new perspective, cloaked in darkness? I loved being awake at night – to have nature to myself in solitude – and now I was alone I knew I would have more time and freedom to explore it. As a naturalist, should I challenge myself to delve into the night and change my perceptions of nature, something I professed to intimately know and love? What would happen if I saw the nightscape as a new land to explore, to interact with new species and their behaviours? What if we stopped clinging to our sense of sight and indulged instead in the sounds and smells of the darkness? There in the garden, a soft charge of electricity pulsed through my body. I looked up at the waxing moon in silence and my thoughts dissolved; for as Antoine de Saint-Exupéry once observed, night is 'when words fade and things come alive'.

My foray into night walks started the summer before when Dave was often away drumming with his band, and I had plenty of time to spare. One evening, after a depressing binge of Louis Theroux documentaries, I found myself longing for the night air rather than surrendering to sleep, so I drove to a little Marilyn nearby called Butser Hill, the highest point on the chalk ridge of the South Downs and one of the darkest areas in the National Park. A Marilyn is a name given to mountains or hills at least 150 metres high, coined as a light-hearted contrast to the Scottish term Munro used to describe a mountain that is at least 3,000ft (or 914 metres) high. Butser takes its name from the Old English *Bryttes Oran*, meaning a flat-topped hill or steep slope, and it's also where Del Boy's hang-gliding scene was filmed in *Only Fools and Horses*. From the summit, there's a 360-degree panorama of the landscape from Portsmouth to Winchester, and over the years it has become a popular observatory for stargazers.

When I was younger, we drove up to Butser to fly kites. Our golden retriever Murphy would lollop about in the meadowsweet blossom, and we bought ice cream from the roundhouse-shaped visitor centre. The area is famous for Bronze and Iron Age archaeology, and the centre paid homage to the types of houses that filled the landscape here 2,000 years ago. Spread across the peak of the hill were remnants of Iron Age ditches, earthworks and lynchets, glimpses into the past that were now not much more than shadows in the earth, darkened patches of soil showering where a timber frame once stood or where animals were enclosed. I would later go on to work at Butser Ancient Farm, an experimental archaeology site nearby that specialises in reconstructing roundhouses and other prehistoric homes to understand how our ancestors lived.

My years at the farm were mad and fruitful. In 2002, before I worked there, the farm was featured on a Discovery Channel documentary called *Rebuilding the Past*, in which a group of archaeologists decided to build the first new Roman villa in Britain for over 1,000 years. The farm is now well known for the reconstruction of ancient buildings based on their archaeological footprints, including the Neolithic, Iron Age, Roman and Anglo-Saxon eras. Since the project was first launched in 1972, it has also become a sanctuary for the weird and wonderful, drawing in visitors and volunteers from around the world who nurture a love for an older, forgotten way of living. Take a glimpse on a warm afternoon, and you'll find Jim casting molten bronze into axe heads, Vivian building a wooden shave horse, Will displaying his collection of Roman weaponry, the education team teaching children to weave wattle fences, or Karen lighting a fire on the roundhouse floor. I once spent a hot morning applying sun cream to our Saddleback pigs, whose pink belly stripes are vulnerable to sunburn.

The farm is also where I developed my love for goats. Specialising in rare-breed livestock, Butser usually has a flock of Manx Loaghtan sheep, a few fat pigs and a herd of English goats, an animal that was first domesticated thousands of years ago in the Middle East. Intelligent, kind and mischievous, goats are underrated creatures, and during my time at the farm, I formed a bond with our girls, particularly Yarrow, Bella, Sorrel and Áine. I helped them give birth, brushed their coats, fed them banana skins, took them to win rosettes at the Singleton Rare Breeds Show, and milked them by hand to make fresh cheese. Goats are magical companions, even if just for a warm hello on a rainy Tuesday afternoon.

Working at the farm gave me a taste of an older, simpler way of life. After our blacksmith Joe showed me how to wield fire and iron, I forged my own butter knife, and woodworkers Darren and Viv taught me to carve wood so I could make a butter paddle to match. Our costume collection grew more extravagant every year so that one summer I stood in the sunshine dressed as a Roman citizen. It is a beautiful place filled with bizarre and wonderful opportunities, but some of the best memories I have were not from crafting, livestock or archaeological experiments; they were made during the long nights I spent beside the fire of the Little Woodbury, our largest Iron Age roundhouse.

The Woodbury is based on a real roundhouse discovered near Salisbury in 1919. Before its excavation, the site was first noted as a crop mark on aerial photographs, but its significance wasn't fully realised until the German archaeologist Gerhard Bersu started excavating in 1938 after he was forced out of his profession by Nazis and emigrated to Britain. Bersu's work disproved the theory that Iron Age people lived in holes in the ground (what we now know to be storage pits), and the Little Woodbury was one of the first roundhouse excavations to shed light on how they truly lived. The building measures approximately 15 metres across and 9 metres high, with a thatched roof, wattle and daub walls, and a chalk floor lined with deerskins for fireside comfort.

I was running an event late one evening when a few of us decided to stay over at the farm rather than deal with the hassle of locking up in the dark. We would be sleeping in the Little Woodbury with a fire to keep us warm and tightly bound hay bales for beds. Before heading in, I offered to walk a loop of the site to check for escaped sheep or lingering visitors and wandered off to the car park alone. The sky was completely clear, the path lit by stars, and the moon almost full, a milky-white orb suspended above me like a big French brie. I walked to the car park where there was nothing to see even in daylight, but beyond that, the farm boundary blurred into the edge of the landowners' hillside forest next door. Dark conifers stood militantly along the land, and as the slope climbed to the top of the peak, fleets of elder trees and other deciduous species were entwined between them; by June the hillside would be a whispering mob of sweet pines and elderflowers frothing in the sunlight. One spring I even heard a cuckoo.

Pesticides and rodenticides are not part of ancient farm maintenance, and the abundance of small mammals, voles, mice and rats had resulted in a healthy population of tawny owls and barn owls as well as buzzards, red kites, kestrels and the occasional hobby. Nearby farmers put up owl boxes on their land, which – combined with the natural materials, gaps and sheltered warmth of our houses – meant that the site was alive at night with shouting owls calling out to the dark countryside. Our Saxon longhouse had been popular with the resident barn owl, who dropped in on winter nights to digest and regurgitate its prey. In the morning we would find shining black pellets full of jaw bones and tiny teeth. I spent many weeks trying to capture it on my trail camera, climbing onto the roof rafters to strap it onto the oak beams, adjusting it every few days to target different perches the owl might favour. At last, I caught one 10-second shot of her sitting roundly on the timber, shimmying her head like a baby snake before the clip ended and she was gone.

The owls weren't calling tonight, but as I walked back towards the farm, I heard rustling in the trees – perhaps an owl preparing to leap out or a pigeon in a restless shuffle. I continued my slow loop around the site. Past the turfed roof of the Neolithic house decorated with burgundy wall paintings that swayed and shimmered in the light of the fire (copies from an excavation somewhere towards Turkey). Past the Roman villa with its cool, white walls and its herb garden blossoming outside. Past the goats who, I discovered after they accidentally misaligned the trail camera onto their paddock one evening, don't go to sleep but continue to roam about all night, wandering intermittently across the camera with white eyes glowing.

For the first time, I saw the farm as an alien landscape; the buildings and pathways that seemed so familiar to me in the day were transformed, and I was thrown into a new world, anchored only by that intoxicating aroma of woodsmoke permanently infused in the air. The farm was cloaked in shadow, concealing the fences and landmarks that defined it, and it was a different place, strange and inscrutable. With one last glance at the moon, I retreated to the roundhouse and pushed the doors closed behind me, shutting out the darkness until sunrise.

Inside, the fire was dancing, and sparks drifted up towards the roof, extinguished by the lack of oxygen. Archaeological experiments completed here in the past included one to see whether a roundhouse roof was built with a hole in the middle to let out smoke. It wasn't, which the team realised fairly quickly before calling the fire engine. With a hole in the top, the smoke and flames were caught in an updraft, and the roof caught fire. They have since proved that smoke filters out through small holes in the thatch, not only clearing the air inside but also keeping insects at bay and deterring birds from pecking the straw. It does not, however, prevent them from nesting in the roof, particularly one as high as this, and a pair of barn swallows arrives from Africa every year and builds a nest in the centre of the roof, directly over the fire. In spring they collect mud and fibrous grasses to make their nest and brood a clutch of chicks, which fledge successfully in time for a second attempt later in the summer. If you sit quietly on a warm afternoon, the adults swoop back and forth through the hole at the top of the door, bringing food to their young in a slow silence broken only by the sweep of air against wing and the *seep* of a hungry chick.

The fire cast a papaya glow over our faces, and I felt strands of fur between my fingers, dusty remains of the old deerskin that separated me from the floor. To us, a tribe of modern people, it was a luxury, a novelty to escape the twenty-first century and retreat to this ancient world, surrounded for one night by the calm of the countryside. I imagined the lives of our ancestors, without modern medicine, electricity or dominion over the earth, but instead nurturing a connection with the landscape, observing the cycles and rhythms of nature and living out their place in the ecosystem. For thousands of years we were prey and predator, surviving as best we could within our ecological corner. I wondered at what point we rose up and started to carve away the land for profit, exploiting other humans and animals to create systems and empires that would lead us further and further away from a natural existence.

I imagined the original Woodbury roundhouse, how it would have stood around 2,000 years ago, built by a community of people who could twist plant fibres into rope, weave wattle fences from hazel, mix together the soil, straw and dung needed to daub the walls together, and thatch a golden roof in the sunlight. The fire would have always been ablaze; the continuous presence of smoke in their lungs may have affected their health, but living an active life in the open air would've tipped back the balance. At night, just as we slept around the firepit, they would have eaten hot venison stew, wrapped in furs and woven cloth dyed with the plants growing outside. Bundled up inside this warm and sheltered space, stories would have unfurled across the flames, passed down through generations from grandmother to father to daughter. Before Christianity spread through Britain, these stories would have been full of natural symbols, tales of the forests and mountains, sun gods and mythological creatures. Historians might now call them pagan, but they weren't boxed in by religious labels; they respected the natural world and wove its power into their lives.

We keep some stories alive in our seasonal celebrations, but most have been lost in the threads of time. Only swallows could remember how these people existed, the birds sweeping over their fireside days and nights for thousands of years, raising their own broods, their own tribes in a roundhouse roof. Would the swallows still be nesting in 1,000 years' time?

In June I came at dusk to Butser Hill to listen to the skylarks just across the dual carriageway from the farm. It's a chalkland habitat and nature reserve, home to many rare butterflies including the Duke of Burgundy, chalkhill blue and silver-spotted skipper, as well as lichens, wild orchids and meadow pipits. During the breeding season, skylarks sing from early dawn until late evening to attract a mate. The energy required for this is enormous, but by singing his best song, a male can indicate his physical fitness and sexual availability to his suitors. All summer they sang under the bright evening sky, and I lay in the grass to listen as the sun dipped below the horizon.

It was now autumn, and the roundhouse visitor centre was closed until spring. It sat in silence beside the car park, pointed roof silhouetted against the sky like a witch's hat. It was not yet six o'clock in the evening, but already the air was closing in, and darkness was settling, the birds quiet, asleep. I left my car by the gate to avoid being locked in later and headed down along the pathway that cuts into the side of the hill. Despite the glow of Portsmouth to the west, there was true darkness here. Conscious of the steep slope to my left that tumbled down to the shadowed abyss known locally as Grandfather's Bottom, I stepped carefully along the track to avoid the brambles sprawling out like serpents. The slopes of Butser are perilous, but come winter they would be filled with families on plastic sledges, whizzing down the hill on a layer of snow as the sheep watched, alarmed, from afar.

At the summit of the hill, a large aerial mast stood next to the trig point we once walked to to induce my sister's overdue baby (now Meredith). This mast, while undeniably ugly, is the unofficial marker by which the locals determine where they are. It can be seen from almost anywhere in the surrounding area, and we rotate around it like a galvanised sundial. That night I used the mast to navigate the hillside, its great barred form outlined against the sky.

I reached a small copse of hawthorn and elder trees where I knew a hidden path tunnelled through to the top of the hill, but the gloaming had crushed all the light from inside, and I realised it would be a blind ascent through the trees. With caution, I climbed through the dark until, emerging from the copse, the track undulated through the grass in loping waves bordered by heaps of cow dung. The trees surrounding the car park and paths began to ebb back to the edges of the hill as I climbed, until I was alone, exposed beneath the stars, a wasp on a yoga ball. There were no clouds, just a pale glow from the distant city and a spread of constellations and satellites glittering across the sky.

I followed a path across the hill, so the dual carriageway and city lights disappeared behind me. It was almost dark now, and I made a note to keep in sight of the pylon for my return journey. Down slightly, and over to the mounds and tumuli on the northern edge. The grass was soft underfoot here, stretched over the milky chalkland, their rich soils bringing so much biodiversity to the South Downs. Eventually, I came to the spot I was searching for: a polo-shaped mound on the slope I had discovered in daylight a few weeks before, facing away from the sea and towards the rural landscape of Hampshire and Sussex. I climbed into the mound and lay down with my back propped against the soil, dirt and dock leaves pressed against the palms of my hands.

These mounds were remnants of Bronze Age barrows and Iron Age ditches, echoes of ancient life when the trees were first chopped down to make room for wattle houses and sheep grazing. The sheep still belonged to the landscape, but the houses moved down over the centuries until they settled in the towns and villages which now lay below me in a sea of sparkling lights. Electricity is a captivating force, but it is nothing compared to a glance at the sky when the coercive lights of modern life make way for the real thing: a shimmering assembly of stars, burning through the night like a thousand gilded rambutans.

They were burning on Butser that night, raw and frozen in the infinite space above us that always makes me feel so reassuringly irrelevant. There is something satisfying about the idea that, while we fritter away the days worrying about office politics and calorie counts and luggage allowances, the natural world is wholly disinterested. The sun will rise, and the moon will wane, and we have no power over it whatsoever. Snuggled up in my ski jacket and gloves, I could feel the cold permeating my cheeks and nose, but I was happy. The galaxy was performing just for me tonight. Squashed into my polo-shaped tumulus, disrespecting some ancient soul buried beneath me, I could see constellations in every direction, each telling a tale on the midnight stage.

The night sky starts with Orion with his belt of three bright stars: Alnitak, Alnilam and Mintaka, which sounds like a spicy dip. On winter nights he's found in the northern hemisphere, although his reputation as the Hunter changes between cultures. In medieval Arabic astronomy he is called *al-jabbar* or the Giant, and in Afrikaans, the belt stars are the *Drie Susters* or Three Sisters. For the Ojibwa Native Americans, he is known as *Kabibona'kan*, the Winter Maker, while in Greek mythology he is the son of the ocean god Poseidon. One day he travelled to Crete to hunt with the goddess Artemis and her mother Leto, but after threatening to slay every beast on earth, Mother Nature sent a giant scorpion to kill him instead. The Greek goddesses begged Zeus to commemorate Orion's death by placing him in the heavens among the stars, together with the scorpion that destroyed him.

My star sign is Capricorn, although I haven't paid attention to it since I stopped reading *Top of the Pops* magazine in 2002. Capricorn is a water goat who lives in the unspoiled wilderness known as Arcadia. He earned his place in the heavens by helping Zeus to fight the Titans, but to escape one of the monsters he was forced to jump into the Nile, transforming half his body into a fish. Today the constellation Capricornus is found in the aquatic end of the night sky with Pisces, Cetus and Aquarius, looking like a squashed slice of pizza.

Draco the serpent dragon lies close to Ursa Minor and the North Star, associated with the dragon Ladon from Greek mythology. Tasked with guarding the golden apples of the Hesperides, Ladon was killed by Heracles who stole the apples as one of his 12 labours. In Arabic astronomy, however, the head of Draco is not a dragon at all, but the Mother Camels. Two hyenas – the stars Eta Draconis and Zeta Draconis – are attacking a baby camel that is protected by four female camels owned by a group of nomads camping nearby. Not as dramatic as a dragon, but the changing stories behind our constellations are a way to explore how different cultures have understood their place in the world. I like to imagine two people looking at the stars from two different points on the globe, projecting their own rich histories onto the night sky. Of course, an essential piece of dragon-based folklore comes from the 1996 film *Dragonheart*, an icon of modern cinema that critic Roger Ebert claimed 'no reasonable person over the age of 12 would ever be able to take seriously'. Here we learn that both the Greek and Arabic cultures were wrong, and the constellation Draco is actually dragon heaven where all dragons go once they have upheld their ancient oath to protect mankind. It's what the silver screen was made for.

For an hour I sat by the edge of the hill and watched the stars until I could no longer feel my fingers. Then, with no choice but to head home, I stood up and left the northern edge of the hill to walk back towards the car park. The problem was that the horizon was no longer glazed with the dying rays of the sun, and the pylon's silhouette had vanished into the shadows while I'd been walking. I had no idea which direction to take.

'Hmm.'

For a second I glanced back at the stars to see if I could use them for navigation, before remembering I'm not a fifteenth-century sailor but a helpless twenty-something dependent on Google for everything. I had no option but to wander off in the vague direction I appeared from and to carefully pick my way across the landscape in search of something familiar.

I walked back to the summit and headed towards Orion's fateful acquaintance Scorpio, not knowing if this was useful but enjoying the view nonetheless. After a while I heard something move to my right; the stars were just bright enough to illuminate the ground, but now what was walking beside me? It sounded like a mammal, but it scampered off with an enviously sharp sense of direction before I could find my phone to use as a torch. Perhaps a fox or a deer; maybe a descendant of the badger whose skull I had discovered in the elder trees nearby when I was younger.

It's a strange thing to be part of my generation and to have returned home, to the place where I spent my entire childhood and teenage years, by the age of 22. Four years later, most of my friends were still living in London, Australia, Thailand, Europe, spending their days exploring and pursuing careers that couldn't really be attempted in a pretty market town boasting of its history in the wool trade. For those of us who went to university – who couldn't wait to escape at 18, live in shared accommodation and eat toast for every meal – only four years away felt too little. I've found inspiration in the lives of my friends living around the world, but I love being at home: watching how a familiar landscape has changed in the short time I've been alive, remembering the highs and lows of each season as they passed, the way the lake froze every few winters, and how tall the chestnut tree grows on Sheet village green while we drank cider in the pub garden.

As I continued my walk along the hill, I realised the pylon had loomed out of the darkness beside me. There she was, the Aluminium Lady, come to pull me out of my dark puzzle. From here, I could align her with the glow of Portsmouth on the horizon and use the fizzing, orange serpent of the A3 like a compass to guide me back to the car. I was never very good at orienteering; when my turn came to navigate our Duke of Edinburgh group across the Isle of Wight, we ended up lost in the rain. This evening, I wondered if my ancestors would be ashamed by my use of a metal tower and dual carriageway to lead me home rather than the natural maps of the landscape.

But the road did guide me, and in the cold night, I climbed wearily into my car to drive back to the warm nest of home and a mint tea.

JAMAICA
INN
TIF

CHAPTER TWO

Ghost Stories

As children we are taught the night is a time for rest, not for play. We have a bedtime to adhere to, a nightlight to ward off our fear of the dark. I couldn't sleep without the clunk of a cassette tape, the soothing voice of Stephen Fry reciting the latest *Harry Potter* book. We are told stories of ghosts and goblins, of werewolves and creatures that lurk in the darkness so that even if we don't fear the night, from a young age, we are taught to be wary of everything about it. Many of the stories we have told ourselves over the years reflect this: they are attempts to know the unknown or to understand our place in the universe. When we're unable to make sense of something we cannot control, we engage our most powerful asset instead: the human imagination. To examine the literature we have produced as a species is to gain insight into our hopes and fears and into the way we have connected with unknown landscapes and ideas as our societies have changed.

I remember finding a badger skull on Butser Hill when I was seven or eight. I liked collecting bones and feathers and other bits of wild debris valued by children and stored them in a plastic box under my bed. The skull was shaped like a diamond with a hard crest along the top, and I remember clearing moss and soil from between the eye sockets. I was never bothered by skulls and bones, despite understanding they were dead things, but I was frightened of taxidermy. Perhaps it was because a skull is almost a new creature in itself, a thing of fascination, an anatomical artefact. Taxidermy looks so much like the living thing it once was, but without that glow that makes living things so beautiful. It reminds us how thin the veil is between life and death – something I avoid thinking about at all costs.

Around the same time that I found my badger skull, we visited Jamaica Inn on Bodmin Moor during a family holiday in Cornwall. It's a haunting old shack, built in 1750 as a coaching house for changing horses and made famous by Daphne du Maurier's classic novel about a group of murderous shipwreckers who drown sailors and loot their cargo. Du Maurier's story was published in 1930 after she had been lost on the moors when out riding her horse. Swamped in fog, she sought refuge at the inn, where she was entertained by the local rector with tales of ghosts and smugglers. While many associate the name with the Caribbean sweet rum smuggled and stored at the inn, it actually came from the local Trelawney family of landowners, who served as governors of Jamaica in the eighteenth century.

It was nightfall when we arrived after a warm day on the north Cornish coast. By then it had grown grey with fog, and we drove through the moor in silence even our empty bellies couldn't break. The roadside was speckled with lumps of granite, but the mist had closed off the rest of the landscape, the steep tors fading away into nothing. After a while we found the inn enclosed in a stone courtyard with a blue-and-gold macaw on a sign and an anchor slumped outside. I stepped out of the car and gazed down the road we had just emerged from. The sky was slate grey, the moon obscured by swathes of cloud that brought a humidity down on our shoulders. I remember tasting salt on the air, imagining gangs of smugglers hiding behind the rocks and waiting for a clear road to commit some dastardly crime. My cousin Calum and I had once raided the drinks cabinet at home when the adults were distracted, so I knew what rum tasted like – appalling – but I had heard it was a favourite tipple with smugglers on the go. Was there rum in Jamaica Inn? Were they preparing for a killing spree? Would I ever make it home to watch the latest episode of *Noah's Island*? Dark thoughts for a dark night.

Inside the inn, we were told the story of a stranger who was passing through Bodmin Moor one misty night. He was standing at the bar, drinking a tankard of ale when he was called outside. He abandoned his drink and stepped out into the darkness. The next morning his corpse was discovered on the moor, but nobody knew how he had been killed or who had summoned him outside to his death. Today, the residents swore they could sometimes hear footsteps in the passageway to the bar, believed to be the dead man's spirit returning to finish his ale. Visitors had also occasionally noticed a stranger sitting on the stone wall outside, who neither moved nor responded to their greetings.

Our evening passed without murders or lootings, but that didn't distract me from what else lay within that fateful place. At that time, it was home to the Museum of Curious Taxidermy, a collection of whimsical tableaux by the Victorian amateur taxidermist Walter Potter. He was famous for his anthropomorphic scenes of stuffed animals mimicking human life, which were displayed at his museum in Bramber, Sussex during his lifetime. Titles like 'The Kittens' Wedding' and 'Guinea Pigs' Cricket Match' are easy to visualise, but he also created extraordinary pieces like 'The Death and Burial of Cock Robin', which included 98 different species of British birds and became the centrepiece of the collection. Google it, I dare you.

In 1984 the collection was bought by Jamaica Inn and attracted around 30,000 visitors a year, including my parents, sisters and me in the late 1990s. Being an amateur, Potter's taxidermy was so contorted and anatomically incorrect that the results were fascinatingly horrible, and they have been etched on my mind ever since. When the collection was put up for auction in 2003, artist Damien Hirst wrote to the *Guardian* about Potter's complete lack of knowledge on anatomy and musculature, claiming: 'Some of the taxidermy is terrible – there's a kingfisher that looks nothing like a kingfisher.' He offered to buy the whole collection for

£1 million, but his bid was rejected. Instead, the pieces sold for just £500,000 and the collection was dispersed.

We spent that night eating dinner in the inn, surrounded by tales of ghosts and smugglers and these macabre pieces of natural history. I remember yellow teeth emerging from shrunken jaws, stuffed rodents in miniature smocks and dresses, squirrels drinking from dust-smeared teacups that would live forever behind a glass display, cold and untouched by human hands. I had not been so glad to leave a place since visiting the Brading Waxworks museum on the Isle of Wight, although after seeing my distress, my dad suggested the animals had 'probably' died of natural causes. I believe him to this day.

Daphne du Maurier's *Jamaica Inn* is one of many examples in literature where the night is used to shape the narrative. It isn't a coincidence that the shipwreckers work under cover of darkness, although they need the night to lure in a lost ship with false lights.

In this story the night is an accomplice to the wreckers, shielding their crimes from the rest of the world and drawing the ships in to their perilous end. The night is so often associated with evil in literature: Dracula emerges at dusk to feed on the blood of young virgins, Othello strangles Desdemona in her bed, and Alec d'Urberville forces himself on Tess in the depths of a dark forest. But for Daphne du Maurier, the night is not only a force for darkness. At the end of her most famous novel, *Rebecca*, the unnamed protagonist drives with her husband towards Manderley, the house haunted by the presence of her husband's first wife and the horrors of the past. As they draw closer, they realise something isn't right – Manderley is ablaze in the night, burning to the ground under the blood-red light of sunrise.

In this story the night is a cleansing presence, burning away the relics of the past to clear a path for the characters to move forwards; a sign of freedom, of restoration. Most of du Maurier's novels were carved into the Cornish landscape, a place that brought liberation into her own life. She once wrote that in Cornwall she found 'the freedom I desired, long sought for, not yet known. Freedom to write, to walk, to wander. Freedom to climb hills, to pull a boat, to be alone.' We witness this freedom in the closing pages of *Rebecca* burning through the night like a cold cremation.

When I was studying for my A-levels at college, I spent my weekend working between the delicatessen counter in Waitrose and the tearoom at Uppark House in West Sussex. They were both excellent, food-orientated jobs and, having just left secondary school and given up compulsory PE, I inevitably grew chubbier. Despite this, I loved working with food and embraced every opportunity to try new things, to taste pungent cheeses from southern Italy or cut up sticky baklava on trays that never quite made it to the sample table.

Working at Uppark House was an enjoyable job for a 17-year-old, although I was less interested in the seventeenth-century architecture than in the fact I could have a laugh with the other girls, drinking lemonade and taking home bulging bags of leftover scones at the end of my shift. Mum was delighted by this and started getting annoyed if leftovers failed to appear. In the summer we would sell melting ice creams in the sun and trap wasps in jam jars to appease customers, which I now regret.

The house is a National Trust property, which means the hygiene standards are high and the bins are located 3,000km from the tearoom. A slight exaggeration, but for a teenager at the end of her shift, tired and dreaming of looted scones, the 10-minute wheelbarrow-walk to the bins was annoying. One night, late in the season when it was already dark by five o'clock, I wheeled my barrowful of bottles and teabags down the long drive and round to the courtyard where the bins were kept. They were huge things with the sort of lids that seem designed not to be opened without considerable effort, and each night I divided up my barrow-load into general waste, clean cardboard, compost, and the satisfying crash of the glass bank.

That evening I deposited my goods and had just turned to start back when I stopped for a moment. Ahead of me, the road stretched on towards the gatehouse, but behind that the estate was thick with trees. Towering oaks had started scattering acorns for the jays to collect through the autumn, and earlier at the top of the conifers, I had seen goldcrests on the smallest branches, jigging about like hot popcorn. Summer was fading, and the air was filled with the welcome aroma of decay that swims out of forests on September evenings.

Above me, a tawny owl was calling. Tawnies are surprisingly small birds of prey, about the size of a healthy woodpigeon, their reddish-chestnut plumage flecked with cream. Somewhere in the canopy, I could hear the shaky *hoo-ooo-oooo* of a male, which I've always thought sounds like he's trying to pick a fight with someone much bigger than he is. I put my wheelbarrow down and stood, listening to the owl offer his greetings to the night, and recalled one of my favourite poems by Wordsworth, in which a young child communicates with the owls at dusk:

At evening, when the earliest stars began
To move along the edges of the hills,
Rising or setting, would he stand alone,
Beneath the trees, or by the glimmering lake;
And there, with fingers interwoven, both hands
Pressed closely palm to palm and to his mouth
Uplifted, he, as through an instrument,
Blew mimic hootings to the silent owls,
That they might answer him. —And they would shout
Across the watery vale, and shout again,
Responsive to his call, —with quivering peals,

And long halloos, and screams, and echoes loud
Redoubled and redoubled; concourse wild
Of jocund din!

Suddenly, hidden in the trees far in the distance, a second call cried out in the gloaming, like Wordsworth's owls responding to the boy. A female, strong and sharp, was calling to the male, and he was responding in turn, sounding more assured than when he had been alone. The famous *too-wit too-woo* sound of the tawny owl is actually two owls communicating; the female utters a shrill *too-wit* in search of her lover, and the male answers back with a reassuring *too-woo*.

For a while I listened to the owls calling to each other, wondering if I was witnessing some magical avian romance unfolding or if it was more of a *Did-you-have-a-good-day?* conversation. As night closed in and spread through the grounds, any velvety fragments of blackbird song lingering in the trees faded away, and all was quiet except for the haunting calls of the owls. Except it wasn't haunting. We weren't in a ghost story, the owls and I, and the forest wasn't a gruesome backdrop for some terrible tale. It was a rustling, moving world, hidden beneath the darkness but, like everything in nature, it was neither good nor evil. Out there in the trees, mice were eating roots, insects were climbing along rotten bark, a stoat was eating the remains of a rabbit, and everything was synchronised with the rhythms of nature. I was finishing work, and the owls were courting through the trees, all of us waiting for the day to close and the next stage of life to begin.

A few years after leaving Uppark, I discovered the writer H.G. Wells had also spent time there as a boy when his mother was Uppark's housekeeper for 13 years. I studied Wells on my undergraduate degree, and his science-fiction works from the 1890s are among my favourite stories of all time. Sadly, I also read his novel *Ann Veronica* while writing my thesis on the politics of women's suffrage in literature, and my loyalty to Wells was tested. A young woman casts off the shackles of social convention to pursue politics and equality, only to meet a man, fall pregnant and give it all up for motherhood. I later read the theory that this story was based on one of Wells' own failed relationships, with the resolution providing the fantasy ending he so wanted in real life. Ah, men.

Despite his later failings, I've remained loyal to Wells simply for the raw storytelling power behind his earlier science-fiction. *The Island of Dr Moreau* is horrifying reading, and in true *Frankenstein* fashion, illustrates the immorality of the actions we perform, even today, in the name of science and progress. *The Time Machine* was thought to be inspired by Wells' childhood spent at Uppark, and it wasn't difficult to imagine the tunnel that leads from the main house to the kitchen – now the tearoom where I served scones and soup – as the perfect dwelling for Morlocks to skulk while the weak and ditsy Eloi scamper through the ancient relics above ground. But it was his 1897 tale *The War of the Worlds* that stayed with me most – the classic story of aliens invading suburban London, obliterating the fortress of the British Empire.

Of all the stories associated with the night sky, alien invasions have the broadest scope. We can tell ourselves anything we like about the horrors of planet earth, what hides in the forest or our own basement, but it is all enclosed within the known boundaries of our world. To look into space and imagine other universes is to imagine the infinite because there is an infinite amount that we don't know. With stories like *The War of the Worlds*, we are twice as frightened because it casts humanity into oblivion, reducing our presence on our planet to dust. The suburbs of London – seemingly one of the most affluent, secure places on earth at the end of the nineteenth century – are invaded by aliens and the capital is almost destroyed. This ruin is a victory for every human and non-human being that has been destroyed at the hands of our own species, made even more satisfying by the narrator's own early ignorance:

> 'A shell in the pit,' said I, 'if the worst comes to the worst, will kill them all.'
>
> ...
>
> So some respectable dodo in the Mauritius might have lorded it in his nest, and discussed the arrival of that shipful of pitiless sailors in want of animal food. 'We will peck them to death tomorrow, my dear.'

It isn't the end of humanity that appeals to us in these stories, but the potential for change. Was there ever a force so arrogant as the human race? Even today, blessed with hindsight, history books, education, ethics and longevity, most of us are still under the illusion we are the rightful rulers of the planet, born to exploit other creatures, people and the environment to suit our own short-term ends. How satisfying to open our minds up to other possibilities by reading about aliens invading earth and putting us all back in our places. If the British Empire can disappear, if humanity can be swept away like breadcrumbs, how many other traditions and institutions can be reworked, renovated and changed for the better? Behind the horror of these stories lies hope; it's the antithesis of the *Daily Mail*.

Aside from its compelling narrative, it was the aliens themselves that drew me into *The War of the Worlds*. The idea of other worlds, planets and communities away from earth is mesmerising, regardless of whether we will ever meet them ourselves. To stand outside, look up at the stars, and imagine another being doing precisely the same thing somewhere else is reassuring. The earth is a lonely place, vulnerable and tiny, and whenever death frightens me I remember the marine biologist and pioneering environmentalist Rachel

Carson who wrote in *Silent Spring* how we are all connected in nature. We are part of something greater than ourselves, a web of living creatures who live and die and whose atoms move on to be part of something new. The idea of other planets, unreachable but still existing out there somewhere, offers consolation amid the madness of life on earth.

Why are we afraid of the dark? Is it the darkness itself that frightens us or the fear of what it hides? Nyctophobia is the name given to that feeling of dread caused by darkness, originating from the reptilian side of our brains that creates anxiety when faced with rational fears like dangerous animals, heights and thunderstorms. For most of our time on earth darkness has meant danger, leaving us vulnerable and exposed, unable to detect threats nearby. Evolutionarily speaking it was once an advantage to be scared of the dark, but in a world of lightbulbs and burglar alarms, this fear has become more of a nuisance than a benefit.

Unfortunately, fears are hardwired into the brain in three different ways and are difficult to remove. The first is by observing others as a child and learning their fears, which is why it's best not to display a fear of spiders around children. Going by this logic, I'm not sure where my arachnophobia came from – my dad loved picking spiders up and letting them crawl on him, but although I've never wanted them dead I still don't like house spiders. Little ones are fine, and I don't mind daddy-long-legs, tarantulas, money spiders or even those jumpy things. But house spiders are the devil incarnate: brown, hairy, thick-legged monsters. We used to have a forest-green carpet in our old house, and on spidered evenings all you could see was a shadow flickering across the floor like the girl from *The Ring*.

The second way for fear to take hold is through a traumatic experience, like falling off a cliff or being attacked by a bear. The third is through a process called anchoring: you have an unexpected fright, and the brain anchors it in some random object nearby, like being attacked outside an arcade or experiencing trauma while watching a particular TV show. Some fears are common and easily understandable, like heights and sharks, while others might plant seeds in our minds that then cause us to enter fight-or-flight mode whenever we encounter that fear again.

Fortunately I don't have many fears and I'm not an anxious person. I don't want to die, I don't like clowns, and spiders are an inconvenience, but I love flying, dogs, sharks, snakes, spaces of all sizes and talking in public. When it comes to darkness, however, it can go two ways. I like to think I'm rational. I can walk around the flat alone at night and enjoy a solitary stumble back from the pub, but as soon as the thought of something creepy hooks onto my brain I'm an irrational mess. Walking through the countryside, which during the day is my favourite place to be, suddenly becomes hellish, full of ghosts and rapists and all kinds of non-existent threats that make me jitter about like Donald Gennaro before he's eaten by the T-Rex in *Jurassic Park*.

There is something in our core that revolts against darkness. Perhaps because we cannot control it or maybe because we know darkness is the default. Lightbulbs, candles, fire, even the sun – every light source we have is temporary, and when they all go out there will be only blackness, just as Lord Byron wrote at the end of his poem 'Darkness': 'She was the Universe.' One of Byron's rare forays into science-fiction, 'Darkness' was written in 1816, also known as the Year Without a Summer and the same year Mary Shelley devised the idea for *Frankenstein*.

In 1816 the world became locked in a volcanic winter caused by the eruption of Mount Tambora in Indonesia the year before. Thought to be one of the largest eruptions of the last 2,000 years, it caused climate abnormalities, flooding and food shortages across the northern hemisphere, with temperatures decreasing by 0.4 to 0.7°C. Aged just 18, Mary Shelley had travelled with her future husband, Percy, to visit Lord Byron and their friends at the Villa Diodati on the banks of Lake Geneva in Switzerland. The weather was so miserable they were forced to spend most of the summer indoors, and it was during one famous evening, circled around a burning log fire, that the group decided to spend the night reading aloud a collection of German ghost stories from the anthology *Fantasmagoriana*. Inspired by the tales, Byron suggested they each write their own ghost story; Mary wrote in her journal the idea that 'perhaps a corpse would be re-animated' and went on to create *Frankenstein*, a now-iconic text in the literary canon 200 years after publication.

Although Britain was still a predominantly Christian country at this point, philosophers and scientists had started to anonymously declare themselves atheist, questioning the truth behind religion in favour of evidence revealed by a growing interest in the natural sciences. An unknown author, later identified as Matthew Turner, wrote a pamphlet in 1782 suggesting the universe had no need of God to guide it, as the world possessed its own 'energy of nature' that enabled it to constantly move forward and adapt. Natural scientists like Georges-Louis Leclerc and Georges Cuvier had been studying prehistoric remains of mammoths and mastodons, species long extinct that proved the animals on earth – supposedly created by God and unchanged since their creation – were actually subject to changes in their environment and could live or die at Nature's whim rather than God's.

Other Romantic poets like Wordsworth and Coleridge reconciled these changes by suggesting Nature and God were intertwined. Being a generally benevolent force, Nature would only be cruel if it was treated cruelly, epitomised in the lines from Wordsworth's poem 'Tintern Abbey':

> Nature never did betray
> The heart that loved her.

Nonetheless, it is no surprise that when Byron wrote 'Darkness', Europe was gripped by an obsession with the apocalypse. Lacking the knowledge to understand what had caused the Year Without a Summer, all kinds of theories were thrust into the public arena, such as the idea of sunspots reported in the *London Chronicle*:

> The large spots which may now be seen upon the sun's disk have given rise to ridiculous apprehensions and absurd predictions. These spots are said to be the cause of the remarkable and wet weather we have had this Summer; and the increase of these spots is represented to announce a general removal of heat from the globe, the extinction of nature, and the end of the world.

In July of that year, around the same time *Frankenstein* was born, Byron wrote a biblically charged, apocalyptic poem about the end of all life on earth after the sun goes out. Some view it as an attack against the Old Testament visions of Christianity, while some claim it was a product of the existentialist gloom that had been taking hold of society for the last few decades. Either way, as we edge closer to the ecological apocalypse we are warned about on a daily basis, 'Darkness' is more relevant now than ever before. It is a poem of despair, of horror and of the fragility of life on earth:

> I had a dream, which was not all a dream.
> The bright sun was extinguish'd, and the stars

Did wander darkling in the eternal space,
Rayless, and pathless, and the icy earth
Swung blind and blackening in the moonless air;
Morn came and went—and came, and brought no day,
And men forgot their passions in the dread
Of this their desolation; and all hearts
Were chill'd into a selfish prayer for light:
And they did live by watchfires—and the thrones,
The palaces of crowned kings—the huts,
The habitations of all things which dwell,
Were burnt for beacons; cities were consum'd,
And men were gather'd round their blazing homes
To look once more into each other's face;
Happy were those who dwelt within the eye
Of the volcanos, and their mountain-torch:
A fearful hope was all the world contain'd;
Forests were set on fire—but hour by hour
They fell and faded—and the crackling trunks
Extinguish'd with a crash—and all was black.
The brows of men by the despairing light
Wore an unearthly aspect, as by fits
The flashes fell upon them; some lay down
And hid their eyes and wept; and some did rest
Their chins upon their clenched hands, and smil'd;
And others hurried to and fro, and fed
Their funeral piles with fuel, and look'd up
With mad disquietude on the dull sky,
The pall of a past world; and then again
With curses cast them down upon the dust,
And gnash'd their teeth and howl'd: the wild birds shriek'd
And, terrified, did flutter on the ground,
And flap their useless wings; the wildest brutes
Came tame and tremulous; and vipers crawl'd
And twin'd themselves among the multitude,
Hissing, but stingless—they were slain for food.
And War, which for a moment was no more,

Did glut himself again: a meal was bought
With blood, and each sate sullenly apart
Gorging himself in gloom: no love was left;
All earth was but one thought—and that was death
Immediate and inglorious; and the pang
Of famine fed upon all entrails—men
Died, and their bones were tombless as their flesh;
The meagre by the meagre were devour'd,
Even dogs assail'd their masters, all save one,
And he was faithful to a corse, and kept
The birds and beasts and famish'd men at bay,
Till hunger clung them, or the dropping dead
Lur'd their lank jaws; himself sought out no food,
But with a piteous and perpetual moan,
And a quick desolate cry, licking the hand
Which answer'd not with a caress—he died.
The crowd was famish'd by degrees; but two
Of an enormous city did survive,
And they were enemies: they met beside
The dying embers of an altar-place
Where had been heap'd a mass of holy things
For an unholy usage; they rak'd up,
And shivering scrap'd with their cold skeleton hands
The feeble ashes, and their feeble breath
Blew for a little life, and made a flame
Which was a mockery; then they lifted up
Their eyes as it grew lighter, and beheld
Each other's aspects—saw, and shriek'd, and died—
Even of their mutual hideousness they died,
Unknowing who he was upon whose brow
Famine had written Fiend. The world was void,
The populous and the powerful was a lump,
Seasonless, herbless, treeless, manless, lifeless—
A lump of death—a chaos of hard clay.
The rivers, lakes and ocean all stood still,
And nothing stirr'd within their silent depths;

Ships sailorless lay rotting on the sea,
And their masts fell down piecemeal: as they dropp'd
They slept on the abyss without a surge—
The waves were dead; the tides were in their grave,
The moon, their mistress, had expir'd before;
The winds were wither'd in the stagnant air,
And the clouds perish'd; Darkness had no need
Of aid from them—She was the Universe.

Byron captures the fear of darkness and magnifies it; light brings life, warmth and joy, and without it the entire world will fall to nothing. It's the ultimate horror story, and although we don't immediately think of these things when we're frightened by strange noises in the dark, perhaps at the core of our nyctophobia is the idea that at night almost everything we need to survive is in stasis. How relieved our eighteenth-century friends must have been when the world did not end, the darkness passed, the good weather returned – and with it the molten warmth of unimpeded sunlight.

I had booked into a highly recommended hostel and spa in a wild corner of the Black Forest in south-west Germany. It seemed perfect: cheap and cheerful, serving vegetarian food, with beautiful views of the forest outside and plenty of hiking trails to join. Dave and I had been seeing each other for a month by this point and, having booked the trip before we met, I was half-determined to enjoy a weekend alone because I-don't-need-no-man and half-checking my phone every three minutes to see if he'd messaged me. I had used Google translate to book the trip online and had also enlisted the help of Dennis, a student intern at the farm that year who was on an exchange trip from Germany, to help me decipher some basic information. Confused, he told me that the spa was for nudists, to which I laughed arrogantly at his silliness. He must have read it wrong. But now I was here and the truth was staring right at me, complete with a friendly *hello*!

Never one to let testicles get in my way, I exchanged greetings with the naked man in front of me who informed me he was on his way to the steam room. I then walked up to my room, unpacked my things, headed out to the nearest tram stop and spent the rest of the day exploring the city. There was a beautiful medieval minster, coffee shops serving slices of Black Forest gateau, cuckoo clockmakers, and even a plaque on the old city wall that marked the spot where witch-burnings took place in the sixteenth century. Full of cake and clutching the tiny woolly mammoth I'd bought from the nature museum, I eventually found my way back to the hostel later that afternoon. After a quick rest, I decided to step back outside and explore the surrounding territory of the Black Forest into which the hostel was snugly tucked. Evening was starting to settle and there was something about the trees that seemed to call me in. Something in the air; an effervescence in the shadows.

Known as the *Schwarzwald* in German, the Black Forest takes its name from the dark canopy of evergreens that have given it such a sinister reputation. It is believed to be the forest that inspired the German fairy tale 'Hansel and Gretel', in which a young brother and sister are kidnapped by a cannibalistic witch living in a house made of gingerbread. Like most fairy tales that have been Disneyfied for a modern audience, their origins are much darker and grittier than we realise, with children regularly being killed, limbs chopped off, eyes pecked out and villains burned to death – all to discourage young people from straying outside the rules dictated by society. 'Hansel and Gretel' originated from the medieval period of the Great Famine, when some people resorted to abandoning young children and even to cannibalism to keep themselves alive.

Part of a vast area of deciduous woodland that had been growing in central Europe for thousands of years, the forest was almost entirely removed by intensive logging in the nineteenth century and then replanted with a monoculture of spruce and other evergreens. In the 1990s, three hurricanes ripped through the area and caused extensive damage to the trees, but as a result many parts of the forest were left to return to their natural state and the Black Forest is now a healthy mixture of both deciduous and coniferous species. As the valleys within the forest were fairly inaccessible, farmers started making cuckoo clocks from excess timber to boost their income. It is said that the first cuckoo clock was made by a clock master who loved how the chime of church bells told the local villagers what time it was, and that, as the craft developed during the long, dark winter months inside the Black Forest, more and more intricately detailed clocks were designed until it became a competition between the craftsmen.

Behind the hostel grounds, a path opened out into the woods and wound up the hill to the darkness within. The day was nearing its end and the air was thick with dusk, the sun vanished beneath the trees, the sky shifting into crushed indigo. As I stepped into the forest the temperature dropped and I was alone but for the invisible blackbirds calling down to me from a canopy of twisted spruce and fir.

I walked, and the path zigzagged higher and higher, an arboreal labyrinth inhabited by birds and beasts submerged in mouldering timber infested with fungi. Deep in the woods where the light failed and the forest floor swelled with the aroma of rotting earth and decay, I could see why this place had inspired the fairy tales that haunted German folklore. All the while there was a constant murmuring behind me, over my head, the sound of the wilderness whispering in my ear. I listened to the birds singing their symphonies to the night, and once – and only once – I saw another mammal. She crossed the track ahead of me, pausing at its far edge to glance back before disappearing again into the forest, her bushy, rusty fox brush the last thing I saw. There were no other hikers, no cuckoo clockmakers out to forage timber, no witches or students or lost naked sauna-lovers. I was utterly alone in the woods, and for an hour I wandered through the trees, the fragrance of night unfurling on the air.

At last, I found my way back to the hostel after emerging from the forest about a kilometre from where I started. Inside I discovered the spa was open until midnight, so I changed into my swimming costume (pointless), handed over a few euros to borrow a spa robe and wandered over to the entrance. I pushed open the door and found myself in an empty changing room and, ignoring the instinct to shield my body from the world, I undressed and swaddled myself in the robe before walking through another door that led into a stone garden.

It was pretty. In the middle a bright, clear pool was illuminated cornflower blue by the lights beneath its surface, and surrounding the pool were wildflower gardens and several wooden cabins, all of which represented a different stage in the spa process. I tried to decipher the correct order by using a German information board, but there was no way I was going in an icy plunge pool so I wandered over to one of the cabins with *Dampf* carved on the front, reluctantly removed my robe and stepped inside. It was a steam room, and there were two other people inside who were, of course, also naked. One was a young girl around my age, and one was an older man who looked almost asleep.

It felt strange to be naked in front of people I didn't know, but then I realised how unnatural that was. Women spend our lives being told to cover up and hide our bodies away. At secondary school we were told off for our skirts being too short or our (school-regulation) shirts being too see-through, and as we get older our outfits are deemed skimpy, we're told to 'dress our age' and made to feel cheap if we flash too much skin. Unbelievably, even breastfeeding is made to feel 'inappropriate'. It's so ingrained in our minds that we are made to forget how amazing the human body is, particularly the female form which literally produces and nourishes other humans. Every curve and hair and bone is nature at its finest, and to be told to cover up is a sign of weakness in those who command it, frightened of how strong the female body can be. We are the only animal that frowns upon nakedness, taking the energy we could spend on healing the planet and instead using it to shame ourselves and others for the natural form of our bodies.

However, I wasn't used to being naked in front of more than one person at a time, so for the first 20 minutes I felt awkward and hot. But then, as I started to relax, I realised how liberating it was. I could feel hot slabs of wood against my skin, and the other girl and I chatted about Germany and careers and books while the man snoozed on in the steam, and when I finally left the cabin I forgot about my robe hanging on the cabin wall.

Outside the air felt freezing after the heat of the steam room, but I let it linger on my skin and close up all my pores before moving over to the hot tub, and here I was alone, the only guest in a fizzing barrel of clear, warm water. I could hear nothing but the bubbling of water, and my lungs filled with the cold air that had been expelled from the trees earlier in the day; now they slept in the darkness just like us. All around the garden the forest encircled me like a diadem, spiked shadows jutting into the sky, now deepened into midnight black and scattered with a thousand white stars like salt grains. I felt light and free, looking up to the nothingness above my head, the universe empty of life and fear and pain, and for another hour I stayed out in the night alone, all the while gazing into the trees, the sky, the half-hidden wonder of the dark, slumbering forest.

TIF

CHAPTER THREE

Polar Night

Three months had passed since Dave and I broke up, and by the time December arrived I had been through so many emotions I felt exhausted. Except for a few texts here and there, we hadn't spoken since the break-up, and I had been distracting myself with trips away and visits to friends, spending as little time in Hampshire as I possibly could, but I was starting to run out of money and stamina. I'm an expert at avoiding how I genuinely feel. I'd been angry with everybody, never allowing myself to cry or think about what had happened. It was textbook denial, preferring to pretend everything was fine rather than considering we might have made a mistake. Now it was Christmas, and although I had been looking forward to spending time with my family, a new surge of emotion had started to rise up in my belly. With another grunt of defiance, I quashed it and decided to escape one last time before the new year, before the reality of my situation would need to be dealt with.

As a result, I was now crammed into a budget-airline seat on the runway at Gatwick airport, wearing salopettes, leggings, three tops and a ski jacket – all essential Arctic clothing I was unable to cram into my tiny hand-luggage case. Sweltering, I managed to awkwardly remove the jacket, waggling sleeves in my neighbour's face at least twice, and squeezed it onto the floor in front of me. Settling in for the rest of my journey, I started to relax as I drank a fresh coffee and enjoyed a big, jammy pastry. I was flying to the city of Tromsø in northern Norway in the Arctic Circle – an escape from British December where everyone was full of joy, spending time with loved ones. I usually love Christmas, and I wasn't feeling so miserable that I would begrudge other people's happiness, but this year I wanted to escape and decided to spend five days alone at the top of the world where the sun 'shines' for only two hours a day.

Polar night is a phenomenon that occurs inside the Arctic and Antarctic polar circles when the sun doesn't rise for several weeks, and a blue veil of twilight lingers over the land for three months. Afterwards the polar regions experience night and day the way more southerly regions do, but then in summer there is a switch, almost overnight, when the season of the midnight sun begins and the region is cast instead into 24-hour daylight. These extreme shifts throughout the year have triggered research by numerous psychologists who study the effects of Seasonal Affective Disorder, and it had sparked my interest too. I had started to spend more time outdoors at night – going on night walks, listening to migrating birds, looking for foxes – and it had given me a taste for exploring the darkness. But what would it be like to spend almost an entire week in the dark without the reassurance of sunrise each morning? How long would the idea remain enchanting?

I've never been particularly drawn to one season or another, but I am interested in the natural rhythm of the year and, no matter which season it is, I find myself longing for whichever is coming next. By March I am fed up with the cold and damp, then by September I long for fireplaces and thick jumpers again. In Britain, we are subject to such a variety of weathers that, while the summer can be warm and bright, in the depths of midwinter it can seem like the sun barely rises at all. But in Tromsø, located at 69°N and 350km north of the Arctic Circle, the polar night – and the weather and temperatures that come with it – last from November to January. That's three whole months of darkness without a break. I was only visiting Norway for a few days in December, but I was interested to see how the lack of daylight might affect people who live through it each year, and whether, as a few locals had suggested online, one could even learn to embrace it.

It wasn't only the endless night I had come to experience, however. When I was 10 or 11, I read Philip Pullman's *His Dark Materials* trilogy, a work of fiction so raw and shimmering that it has remained my favourite ever since (followed closely by *The Lord of the Rings* and *Jurassic Park*, naturally). As I read and reread this trilogy growing up, I slowly absorbed the compelling ideas within Pullman's writing but, as a child, it was the setting that stuck with me. The bleak, white landscape of Svalbard, the armoured bears, and those Nordic ports smeared with oil and spirits, and then a natural spectacle so enchanting it has waltzed through my mind ever since: the Northern Lights. Also known as the aurora borealis, they are natural light displays that occur in Arctic regions when the magnetosphere is disturbed by solar wind and charged particles collide to release red, yellow, green, blue and violet lights. I had waited my whole life to witness this phenomenon, and now here I was, finally flying through freezing December skies and into the Arctic Circle. As I looked out of the window, I could see snow-covered islands half-hidden in shadow, the frozen landscape of Norway's northern coast.

I touched down in Tromsø airport that evening, and I took the *Flybussen* into the city centre – a small, glittering place. The city of Tromsø is home to around 75,000 people, similar to the population of Cannes in France, but it feels far removed from the chaos of Europe to its south. Disembarking at the end of the bus route, I walked the last few hundred metres along the harbour, where a gang of eider ducks were floating around in the water. It was dark, but it was late so the darkness felt natural. Lights sparkled across the water, glowing against the sheets of ice frozen to the pavement. I was well wrapped up, but at -9°C this was one of the coldest temperatures my body had ever been exposed to, and by the time I reached the hostel my nose hurt. Through the glass door of the reception, I could see a rotund terrier sleeping on the sofa, but I had arrived too late to meet the owners. The security code was written in felt-tip pen on a piece of paper by the door, and I let myself in around the corner to find a snug dormitory, shared kitchen and bathroom.

I love hostels. They're cheap and friendly and full of the ghosts of past guests, immortalised by the culinary delights abandoned on the 'shared-food' shelves or in the happy messages scrawled in guest books. As long as there are a kettle and a hot shower, a hostel offers all you need for a small adventure. I dropped my luggage in the corner by my bottom bunk and waved hello to the Argentinian guy sitting on the bed above me, warning him apologetically, as is my custom, that I may shout in my sleep. In return he shared a few helpful hints about the city with me and pointed me in the direction of the cheapest supermarket. I wandered into the kitchen to poach a cup of coffee, where I met a friendly Taiwanese man named Frank who was travelling around Europe before pursuing a career in car mechanics. We chatted over coffee before Frank invited me to hike with him to a nature reserve called *Prestvannet* ('priest water'). He had heard the sky was unpolluted there and you could often see the Northern Lights. I accepted his invitation, pulled on my snow boots, and a few moments later stepped out of the hostel and back onto the streets of Tromsø.

This was my fourth visit to Norway, but my first to the Arctic Circle. My sister lived in Bergen for more than three years, a beautiful city on Norway's west coast that's surrounded by seven mountains. While she lived there my family would visit every year and we'd go hiking, eat waffles and get drunk in the snow.

In July 2015 we went white-water rafting with a large group of her friends in the Voss fjords, north-east of Bergen – an unforgettable experience, notably when I fell into melted glacier water and almost froze. Southern Norway is one of Europe's rainiest regions so its mountains are malachite green and the air there is fresh and dewy. Within its forests lie boundless carpets of moss-coated mounds that make it feel like you're in a troll kingdom, and in summer the trees are full of the thrushes that, come autumn, will abandon Scandinavia to head further south. The summer landscape was fiercely beautiful; at that time of year there are no aurora displays, but instead the nights are short and people spend as much time outdoors as possible. The summer we went rafting, we also experienced one of the most surreal nights I could have hoped for in such a magical setting: a night in the fjords with the Gudvangen Vikings.

The fjords of western Norway may look like sapphire lakes but they are actually narrow saltwater inlets from the North Sea, formed by glaciers that had cut deep into the land to produce vast, steep mountains and cliffs on either side of the water as they receded. Individually they can seem separate from one another; however, some of Norway's fjords are intertwined, making it possible to sail from one to another and onwards as far as the sea. In the summer, melting glacier water escapes down the mountains and pours from the cliffs in endless, glittering torrents. Norwegian legend has it that the fiddle-playing spirit *Fossegrimen* lives beneath the waterfalls and he will teach others to play the fiddle in return for an offering of food, although if your offering is deemed too small he will only teach you how to tune his fiddle.

After Hollie's friends and I had all thawed from rafting the Voss fjords, we drove east to a remote spot just outside the village of Gudvangen where the Nærøydalselvi river empties into a fjord. Here we set up our tents in the bottom of a vast valley, ate crisps and had a few drinks while we waited for darkness to close in and the night to begin. We were tired from a long, active day in the sun, but it was far from over yet.

My middle sister can be described as many things, but there is no one in the world as fun as her. The Queen of Good Times, Hollie always has a bottle of prosecco cooling in the fridge and can arrange a party at the drop of a hat. You're guaranteed a good time with her, even if you also end up lost with a broken liver. So when she informed the rafting group that we were going to attend a Viking re-enactment festival called *Gudvangen Vikingmarknad* ('Viking Market') in the fjords, we acquiesced. The festival was an annual summer celebration of all things Viking, with battles, concerts, storytelling, archery, games, lectures, food and even a slave market.

We would be arriving on the final night of the festival when Hollie had heard a great celebration for all the re-enactors would be taking place after a long day of Viking activity. The problem was that it was only accessible to re-enactors, so between mouthfuls of crisps and beer, we helped each other transform into Vikings. Using a paper plate, I fashioned a 10-inch shield with a magnificent eagle drawn in biro on the front. One of the group members, Ole, was draped in a coral-coloured blanket, while Hollie and her boyfriend, Haakon, were decorated with random tufts of fur and scraps of plaited leather. The overall effect was pretty weak; fortunately, two Canadians in our group had prepared outfits that were just about good enough to pass as re-enactment garb, and we hoped we might make it inside if the rest of us shuffled in quickly behind them.

It was now ten o'clock in the evening, and the mountains that engulfed us on either side of the valley had disappeared into the darkness. We could hear water flowing down the river, but otherwise the valley was quiet. We took a last swig of beer then left the campsite and wandered down to the Viking market entrance, doing our best to appear as Norwegian as possible. At the gate, Haakon spoke with the attendant, and I hid my terrible shield behind Ole's coral cloak. After a few anxious seconds, we were admitted and stepped forwards into the festival.

We walked through the grounds and found empty stalls that had been filled with trinkets that morning: carved wooden pendants, cow-bone dice and thick woollen blankets, all bought by visitors who had attended the festival earlier in the day. The visitors were gone now, and the place was quiet. The sky loomed over us like dark slate, and the stars shone in a thick mist, smeared across the ether like glitter glue. There was nothing to disturb the silence except the muffled thump of a drum in the near distance. As we drew closer, an oak longhouse appeared out of the gloom, heavy with timber beams and dovetail joints. The walls were solid, and spilling out the door came a vibrant, tangerine glow and the aroma of woodsmoke. On entering, we found a hall full of men, women and children seated in rows along solid wooden benches, their faces lit by the naked flames of candles, and cats and dogs curled at their feet. Everyone wore shaggy furs the colour of heather and oat grains and they were singing in low, slow voices to the irresistible sound of the quartet playing flutes and lyres in the corner.

There was a space at the end of a table, so the group of us sat down and looked around. There were Vikings everywhere – and they looked beautiful. Many of the women had braided their hair into long, twisting vines, and most of the men were bearded and draped in tunics of grey and moss green. We didn't hear anybody speak English, yet we were immediately offered drinking horns and tankards filled with ale and mead. I befriended a handsome Viking sitting next to me, exchanging three or four words of broken Norwegian. Someone else handed me a kitten.

For hours we all drank mead by candlelight, the quartet performing Scandinavian folk songs that brought several people to their feet, swinging between the tables in an intoxicated waltz.

We were giddy with drink. One by one our group broke off and returned to the campsite until only Haakon and I remained with the Nordic marauders. Candles flickered in the early-dawn wind that swept through the mountains and into our timber hall and I felt my eyes closing against the music. Then suddenly the key changed and the players began a new tune. I opened my eyes and looked at Haakon, and in our drunken haze we were overcome by the realisation of what they were playing: it was the *Game of Thrones* theme tune!

The combination of mead and good cheer had heightened our senses, and we exchanged joyful cries at the beauty of a song, already evocative for anyone who watches the show, played on flutes and lyres at a Viking gathering in the crevice of a mountain in Norway. We danced and swayed with the others until eventually, defeated by sleep, I left Haakon in the hall and sauntered back to the campsite. The handsome Viking I'd been chatting to offered to accompany me back to camp and, unable to share more than one or two words, we ambled back in contented silence. Above us, the stars were strung across the sky like pearls, and the moon hung between them, an ivory pendant. I reached the tent and waved goodbye to my new friend. Settling down to a peaceful sleep in a nest of cushions beside my sister, I was awoken ten minutes later by Haakon setting off the car alarm.

Back in Tromsø, I was walking with Frank along Storgata, a busy street that runs from north to south along the western bank of the city. From here we turned right and followed a sequence of curved roads up a steep slope, all the time climbing and gazing up in the hope of spotting a sliver of blue or green. The pavement below our feet was buried under snow and ice, a crunching pathway to the summit that caused us both the occasional wobble, despite my lovely new snow boots. Frank told me about his travels as we walked. He was from Taiwan, an island country situated in the South China Sea between Japan, China and the Philippines, where he had trained as a car mechanic and salesman with the aim of one day running his own factory. With his last days of education behind him, he had wanted to see the world before committing to a career, and Tromsø was his final destination before returning to Oslo for the last flight home. I asked him if he was sad to be leaving, and he nodded but added that he was looking forward to seeing his mother.

We reached a more suburban area of the city, with fewer shops and more wooden houses painted in shades of rich blood red, almond butter and slate grey. Norwegian homes are beautiful. Ikea, although founded in Sweden, has become a beacon of Scandinavian design and functionality, and when my sister first moved to Norway and wanted to decorate her flat with vintage furniture she'd collected from travels in exotic lands, she soon learned this was not the Norwegian fashion. Instead their houses are clean, colourful and *koselig* (cosy), replete with plants, handwoven throws and underfloor heating that is almost obligatory in such a cold country.

After a while, the road cut through a cluster of pine trees so thick that, even in daylight, we could not have seen through the gaps. Through here we would find our nature reserve, according to Frank, so we left the road to descend along a pathway. The path had been completely smothered by snowfall except for a boot-width trail that had been compressed by walkers some time earlier. We crunched along into the forest for a few minutes, the world beyond the trees blocked out by gnarled branches and mounds of snow that looked like iced buns. I wondered how many birds and mammals were still here, and how many had migrated away to milder places, or were hibernating below ground to escape the painful cold of the snow. A forest always feels alive, but I couldn't avoid the feeling that it was quieter than it might be during the warmer months, asleep rather than awake, lacking the vivacity of summer. While the woodlands of Britain are often grey and damp in winter, you can always smell the leaves from autumn decaying underfoot; here there were no smells, only the blank brilliance of snow and ice on every twig, every frozen puddle.

Emerging from the gloom, we found ourselves standing on the bank of a small lake. Frank told me this area had been protected due to the populations of birds that nest here in summer. Enclosed on all sides by a crown of winter trees, the lake stretched out before us in a sea of midnight blue. Above our heads, the cloudless sky looked like an ink stain that has spread all the way to the edge of a sheet of blotting paper, so the only way to tell where the lake finished and the sky began was a cluster of lights on the horizon, shining from a row of houses filled with people sleeping through the night.

I'd heard that more tourists visit the Tower of London than Londoners themselves, and I wondered if something similar applied here. Did the residents of Tromsø still go outside every night the aurora was displaying and watch the lights from beginning to end? Was it a vital part of surviving the polar night, to admire the beauty of the aurora and forget about the winter darkness for a few hours? Or was it something the residents had grown used to, a display for tourists, an ordinary part of everyday life that didn't justify leaving the warmth of home?

The aurora was not displaying, but a group of Spanish travellers were laughing and drinking at the water's edge, their cameras and tripods arranged defiantly in the snow. We approached and asked if they had seen the lights, and they nodded excitedly that *¡Si!* They had! Frank and I decided to wait by the lake to see if the aurora would return. Our new friends shared a beer with us and we all sat on the snow talking for a while until, suddenly, someone shouted and pointed to the sky.

A single ribbon of light had appeared from nowhere in the sky above the lake. I squinted. It was barely visible at first, a flickering serpent waking from sleep, but soon it started to move and grow. The ribbon stretched across the lake until, as the minutes passed, both ends disappeared beneath opposite sides of the horizon, and the light sliced the sky in two. As the light inched across the sky in wandering, waving movements, a sliver of blue and green seemingly without purpose or direction, it reminded me of a worm slowly contracting and extending itself as it moves through the soil. And as the ribbon widened, it seemed to harvest colours from all over the world, reflecting the cerulean waters of the Caribbean sea, the lime greens of sphagnum moss, the electric blue of a cobalt crust fungus, the pearlescent aperture inside a seashell. In that moment, the entire universe seemed to be captured, drifting through the sky before me in a glass thread.

Like those brooding swallows back home in the roundhouse, there was something about this sight that fixed me in the present and reminded me that I am but one of many mortal beings the natural world has outlived. How many years had the Northern Lights been dancing across these skies? How many humans had gazed up at them in wonder? I remembered reading an eighteenth-century account of the aurora when I was researching my Master's thesis on the Nordic landscape writer Knut Hamsun. According to the first volume of the American Philosophical

Society's journal, a correspondent from Pennsylvania recounted his experiences in the Arctic to the board:

> That about half an hour after seven in the evening of January 5, 1769, there was seen at that place, a bright crepusculum, rising out of the North ... At three quarters after eight, it was so light in the Northern hemisphere, that a person, who felt no decay or infirmity of eye-sight, might easily have read a book printed in Double Pica Roman ... At nine o'clock, five columns or pyramids, of a very vivid red, rose perpendicular to the horizon ... During the appearance the air was uncommonly severe and chilling; and, though the Heavens were serene and bespangled with stars, the atmosphere felt damp and heavy.

The composition of the earth and its atmosphere has fluctuated since the Big Bang, but these lights have surely been illuminating our night skies for a few million years at least. For a long time this marvel was known only to the indigenous people who first settled here. How much the lights must have astounded polar explorers when they arrived in the Arctic Circle. Would they have thought them witchcraft? A sign from God? Or perhaps energy to be somehow harvested and sold for profit?

For an hour we all lingered on the shores of the lake and watched the display. Then the ribbon dissolved into blurred swathes of light and the sky became a pool of colour, reaching down to the land where the trees formed silhouettes against their kaleidoscopic backdrop. With no wind, the lake was a black mirror, reflecting the aurora and trees and the stars that shone faintly through the lights. In the distance, I could see the Arctic mountains standing in the darkness. These giants are perhaps only properly appreciated in the summer when they are more than dark shapes against a technicolour skyscape. Not yet ready to give up on the lights and shielded from the cold by my salopettes, I lay down in the snow and drank them in like rum punch. After a while the colours started to fade, the ribbon withdrew, and the sky returned to shadows and stars. Dazed and sleepy, I could feel the cold in my fingers and nose – but I had seen it. A rippling, glistening grand *jeté* from the aurora borealis, it swept through the air and then was gone. Frank and I gathered our things, waved goodbye to our Hispanic friends and started the long walk back to the hostel where warm bunks awaited.

***

I didn't see Frank again; the next morning he left the hostel in the early hours to catch his flight. I decided not to set my alarm and instead allowed myself enough sleep to feel revived for a day exploring the city. On waking at ten o'clock the next morning, I peered out of the window to see a darkened street looking exactly as it had the night before. The streets were quiet with few people and even fewer cars, and when I poked my head out of the door there was nothing to listen to except the crunch of snow underfoot as people walked slowly past. I'm not sure what I had expected of the polar night, but I quickly realised that it was what I had seen from my window: a landscape cloaked in dusk.

In the hostel kitchen, I ate slices of fruit bread with jam that I'd bought from a late-night grocery on the walk home the night before. An auburn-haired young French woman called Nathalie was cleaning the kitchen. She explained she had been working in the hostel for a few months, changing the beds and cleaning in exchange for accommodation. We shared a coffee and chatted for a while about our travels and graduate life. She had given up a career as a psychiatrist after experiencing anxiety brought on by her work and now she was slowly wandering across Europe in the hope of finding a new life. I asked her whether she thought long winter nights had an adverse effect on the people who lived in northern Norway. She said she had wondered the same, but after some time realised it was people who had moved here from further south who suffered the most. Norwegians who had spent their lives at this latitude didn't mind the dark as much because by the end of their summer – the season of the midnight sun – they craved the cosy solitude of winter and welcomed their snow-dusted landscapes again. Nathalie said she had noticed that people who moved here from brighter, warmer climates found the polar night oppressive, even if they were only relocating from a few kilometres south in Bergen or Stavanger. For these newcomers, the polar night could be incredibly hard going.

After breakfast, I stepped out into the quiet streets of Tromsø and looked around at this new landscape. Every surface was heaped with snow except for the roads, which had been salted to ease the passage of 4x4s with thick winter tyres. The streets were peaceful, empty. I walked down into the hub of the city and found coffee shops with high windows decorated with strings of lights and pot plants, anything to keep the blackness of winter at bay. Nobody was in a hurry here – if they tried, they would fall over – so residents half-walked, half-shuffled between their destinations, slow but never lingering. Over everything hung a cloud of evening light, an unremitting greyness that, by the end of my trip, would make me feel tired and confused whenever I stepped outside. I've never been brilliant at waking up in the mornings, no matter how beautiful the birdsong is or how fresh the air, and I rely on the sunlight to pour in and remind me that it's time to get up and do something with my day. Here, although there was plenty to see and do, I never felt revived or full of energy because for five days I couldn't feel the warmth of sunlight on my skin or the brightness of day in my eyes.

It felt like an eternal grogginess, a struggle to kick-start my mind and body.

There was beauty in it too. From 11 o'clock each morning came two hours of what seemed to me like dawn light, lifting the fog of night and allowing a sliver of lucidity into the city. It wasn't sunlight – the sun was still far beyond the horizon – but it was just close enough for a few sunbeams to creep across the mountains and brush the streets with a golden pallor before the afternoon darkness returned. Without direct exposure to even the smallest ray of sun, it never felt warm, but it was an uplifting feeling to see the city emerge out of the shadows for an hour or two.

In a place like this, trying to avoid the night would mean staying inside for three months. It is impossible to escape it. But walking through the streets past shops, rows of houses, doctors and cinemas, I wondered if you have to claim the polar night in order to live through it; you can't hide from it, you have to bring it to life and make it your own. The fairy lights strung between windows were a beacon of hope and happiness in the gloom of Tromsø, and hanging from every surface during my trip were white and gold decorations for Christmas, bright stars and Nordic elk outlined in phosphorus to make them glow in the dark. I crossed the broad river that cuts through the city to visit the *Ishavskatedralen*, or Arctic Cathedral, jutting above the city like a shard of ice. As I walked towards it that day, lines of hooded crows were roosting along the cathedral's sharp edges, making the building seem like a golden reminder to the people of Tromsø that light will always triumph over darkness.

One afternoon I stopped for hot chocolate in one of the cafes along Storgata. Its front walls were made of floor-to-ceiling glass with Nordic inscriptions; inside, raw copper lamps dangled from the ceiling above timber surfaces covered in candles. The cafe was warm and *koselig*, but its rustic aesthetic echoed scenes I had seen when visiting the *Polarmuseet* (Polar Museum) the previous day. The museum houses a collection of artefacts from the Arctic trapping and polar-exploration industries that Norway was built on. I encountered sights that horrified me including several dead polar bears strung from the ceiling, a stuffed replica of a seal about to be clubbed, and an exhibition on the famed *Isbjørnkongen* (Polar Bear King) Henry Rudi, who is believed to have slaughtered 713 bears in his lifetime. I try to stay relatively open-minded and appreciate the complexities of history, but I can honestly say this was one of the most awful places I've ever been.

The Northern Lights are perhaps Tromsø's most recognisable beacon of hope in the darkness. A walk through the town reveals a roaring trade in aurora magnets, postcards and guided tours to the edge of the city to see the lights for yourself, but the residents' connection with the aurora goes beyond tourism.

The aurora is a wild, uncontrollable light in the dark and, while it can't equate to a sunny day in terms of warmth or Vitamin D, it is easy to see why the people of the Arctic Circle have celebrated this natural wonder for millennia. Inuit tribes around the world share myths and folk tales about the Northern Lights and the aurora's connection with a higher power. Some tribes believe the Lights are the spirits of the dead playing football with a walrus head. Interestingly, the Inuit of Nunivak Island in Alaska claim the opposite and instead believe the Lights are walrus spirits playing with a human skull. In Gaelic folklore, the Northern Lights are known as the *Na Fir Chlis*, or the Merry Dancers, and the displays of light are thought to be epic fights among sky warriors; the stories say that blood from wounded sky warriors fell to earth and caused red spots to appear on the bloodstone rocks in the Scottish Hebrides. The Finns named the Lights *revontulet* – for *revon* (fox) and *tulet* (fires) – because of an old folk tale in which an Arctic fox is running so far into the northern lands that his fur brushes the mountains and sparks fly off into the sky, creating the Northern Lights.

My first night with the aurora had given me a taste for more, so I booked onto a group visit to the edges of the city to go 'aurora hunting' by following waves of electromagnetic activity across Troms county in a little minibus. Our group leader and aurora expert, Emiles, was from Latvia. I asked him how he coped in Norway with three months of darkness; he said he spent a lot of time snowshoeing, a recreational activity involving hiking through the snow using specially made shoes.

Emiles had brought snowsuits for anyone not wearing enough layers. After half the group had pulled them on, we all piled into the minibus and drove towards the mountains. The app I had on my phone informed me the conditions were right tonight. Via the app I watched a minute-by-minute update of a map of northern Europe that showed high electromagnetic activity with what looked like a cloud of nuclear waste moving slowly towards Norway. After 20 minutes or so we stopped by the side of the road. The eight of us had been chatting for the entire journey. Most of us were jammed into the middle of the vehicle, so peering through the windows was impossible. As the doors rolled open, out I fell, landing underneath the most majestic sky I had ever seen.

If I were to compare the aurora I'd seen at Prestvannet lake to a freshwater stream with colours bubbling through the ether, this aurora could only be described as a frenzied, blistering river of lava that was ripping the sky into pieces. With the lights of the city far behind us, the air was ablaze with colour. Blues and greens still shone from the core, but along the edges, the burnt pinks and oranges of grapefruit zest and coral, the violet of aubergines and mallow flowers. The sky was open to us all, and our hotchpotch group of travellers from across the globe joined together for one moment to bathe in electromagnetic beauty.

Emiles gathered up a pile of logs from the back of the bus and lit a fire on the icy road, and we watched the ice disintegrate beneath the flames. For the next hour, we took it in turns to wander beneath the aurora, capturing long-exposure photos, observing the ripples and tides of light as it streamed over our heads like rainbows liberated from their geometric constraints. We drank hot chocolate to warm up, huddled by the fire when the cold penetrated too deeply into our bones and sinews. Beneath my ski jacket and multiple layers, I could feel the scratchy heat of my merino jumper against my bare skin as I enjoyed the hot trickle of chocolate working its way to my stomach. A French couple in our group had started dancing to electro swing playing from their phone to keep back the cold. I hopped around the fire with an Australian girl, who explained that the stars looked different on this side of the world.

Eventually, the Lights started to fade and, secretly relieved, one by one we climbed back into the bus and snuggled together for heat. Emiles drove us further still from the city, chasing the electric glow that was now floating away behind the mountains in front of us. Between the wisps of aurora we also watched the stars, which, although not exactly what we had come here to see, were bright and crisp in the Arctic sky. Onwards we went, winding between fjords and jagged rocks sealed over with snow and ice, as we travelled through the night until at last, around two in the morning, the minibus stopped and we all climbed out again. Here there were no houses, no strings of lights in coffee-shop windows. Before us, a midnight blue fjord reached out into nothingness, a mass of salt water cloaked in stars. It was so dark we felt almost blind. Had we driven to the ends of the earth? Not quite, but we had reached the end of the island, marked by a quiet fishing town called Brensholmen just a few kilometres on. We could see the lights of the town gleaming faintly over the peak of a rocky incline, and I wondered how such a community could survive in modern Europe, both geographically and socially cut off from the rest of the country. Emiles told us that fishing and agriculture had been at the heart of this place since the Iron Age, but today the population numbered no more than 300 people. It was a vibrant little corner of Norway, nonetheless. In winter, tourists came here to see the aurora; in summer, visitors on deck of the regular ferry to the nearby island of Senja could watch orca and humpback whales spyhopping in the Norwegian Sea. There are thought to be around 3,000 orca living in the Norwegian and Barents Sea, with pods often working together to herd fish into tight balls to ease their capture. Beluga whales also live in Norwegian waters, as well as seals, sperm whales and polar bears near the coast where sea ice forms.

Brensholmen was our final stop before the long drive back to Tromsø. We stood in sleepy silence and watched the fjord. There, reflected in the water like a shimmering sapphire curtain, a faint cloud of aurora lingered in the sky as if to wave us farewell. It would be the last fragment of Lights I saw before leaving the Arctic and heading back to Britain, and when my tired eyes looked carefully I swear I could almost see a golden city hidden in the aurora's folds, just as Lyra had seen in the Retiring Room at the start of Pullman's *The Northern Lights*. Standing here beneath a Nordic winter sky, it was easy to understand how the Lights had woven their way into so many folk tales, myths and stories. Intangible and unforgettable, I had witnessed for myself how the aurora charges our imaginations with a murmuring, bewitching electricity. Together our group watched and waited until the last glimpse of colour had faded from the air, the final stroke of a paintbrush gone. The sky returned to darkness, a scattering of white stars emitting their own plasmatic radiance once more. Then – cold, stiff and desperate for sleep – we climbed back onto the bus, Emiles pouring us all another steaming mug of hot chocolate, before the engine started and we began the long journey back to Norway's midnight city.

CHAPTER FOUR

Taxus Baccata

Seep-seep.

Seep-seep.

Two birds in unhurried succession have flown over my head.

Seep-seep.

I can't see them because it's 11 o'clock at night, and the thick cloud cover of late autumn means not even the stars can illuminate the garden in which I'm standing. But I know these birds.

Seep-seep.

They have a creamy stripe above their eye and a smear the colour of rusty pumpkin under each wing, and in the darkness they call *seep-seep* to their friends and flockmates, sharing salty tales of their recent odyssey over the dark North Sea.

Redwings are the smallest true thrush in Britain, and while a handful of pairs breed in Scotland each year, most of our population will not arrive on our shores until October or November. From September onwards, loose flocks of thousands of birds gather along the Scandinavian coastlines at dusk, where they breed through the spring and summer months before launching into the sky on a single 800km nocturnal flight across the North Sea and over to Britain. Many will crash and drown in the waves, beaten by the turbulent weather of an ocean nightscape – so why risk such a journey?

Migration is one of nature's greatest challenges for wildlife, but a mesmerising spectacle for those of us waiting at the other end with a pair of binoculars and a cup of tea. It was only in the last century or so that scientists started to grasp the complexities of the process; why and how do these birds travel such great distances? Before the theory of migration was developed, some ornithologists believed birds either hibernated in trees and at the bottom of ponds or transmogrified into something else. For centuries observers thought redstarts turned into robins over winter, and some even believed that barnacle geese hatched from barnacles, before someone hopped over to Greenland and found their nests.

Now we know they migrate, we can start to understand the logic behind it. When birds need more food than their breeding grounds can provide, they look elsewhere. In winter the Scandinavian landscape is bleak and covered in snow, so birds fly to Britain for a slightly milder season where they can find the berries and earthworms they need to survive until spring, when they will return north to breed and the cycle will be complete. Some redwings migrate from the birch forests and thickets of Russia and Iceland, all the way down to Portugal, Greece and even Iran. The ability of these species to travel thousands of kilometres and not flop down dead at the end is astounding, but there is something else about migration that my modern brain had been unable to resolve until I returned from Norway that winter: how do these flocks of birds know to transport themselves at exactly the same time, as one beautiful, breathing mass of life? We know birds communicate effectively and we know all species are driven by their biological needs. But what sequence of events follows for individuals to gather along the edge of the sea, leap together into the night and spend six months in a different landscape?

Instinct. The innate, primitive response buried deep in our bones that causes us to fight or fly. That winter I caught a glimpse, a taste of the same restless instinct that had driven these redwings across the dark ocean and over my head into new lands.

After many weeks living at my sister's house, by Christmas I decided something needed to happen. I couldn't stay here forever, and living in the south-east meant that renting a place on my own would bankrupt me. Having avoided socialising and going into town – desperate to avoid Dave and the torrent of emotions I was leaving unprocessed – I thought it might finally be time for a change, to start again somewhere new and leave my beautiful corner of the South Downs behind. It was an exciting idea, mostly because it was an actual plan with movement involved. A decision was being made. For a while I considered the dark and vibrant tors of Dartmoor, somewhere I had fallen in love with when we spent a long weekend there in the camper van a couple of years ago. At last, though, I settled on the Peak District. Aside from being the birthplace of the Bakewell pudding – a favourite of mine – it was far away but not so far that I couldn't come home in half a day. With a renewed sense of purpose, I spent the last week before Christmas looking for jobs and houses, starting to put together a plan for the new year.

And then, on Boxing Day, I saw Dave.

He had invited me round to collect some paperwork for the Škoda. I spent half the day feeling sick and the other half convincing myself I was totally fine, both of which should have prepared me for the moment when, on entering the flat, my heart felt like it had been punctured. I stayed for an hour or two and we shared tea and teary conversation, and then I left to go back to my sister's. I tried as hard as I could to squash all my feelings down to where they were bubbling up from, like trying to submerge a float at the swimming pool, and the next day I even drove up north to spend some time with friends and distract myself. But after two days, miserable and confused, I cut my trip short, drove back to Hampshire and got in touch with Dave to tell him how I felt, but I soon realised a Prosecco-infused half-argument on New Year's Eve was perhaps not the best way to address our complex emotions.

On 1 January I started running. Since taking up the hobby a couple of years earlier, my sister Chloë had become a running fiend after discovering the physical and mental benefits, and she'd been gently nudging me here and there to give it a go. A natural scoffer of exercise, I resisted for a long time until I was seduced by the 'Run Every Day' January scheme celebrated by various health charities. It's all in the title: you run every day in January, even if it's just five minutes down the road and back. I love a good scheme, something you can share with others and find support on social media, and I'm foolish enough to believe in (and love) New Year's resolutions. So I started to run.

I didn't start pushing myself too early because I wanted to enjoy it, so my first excursion was a treacle-speed 2.5km to the corner shop and back. And dare I say it, I actually *liked* it. It was the first day of January so I was, of course, hungover and still processing what had happened the night before, but my tiny run filled me with fresh air and brought me slightly back to life. The next day I hopped a little further, and by the end of the Christmas holiday I had even bought a new pair of trainers to cope with the 5km ordeal I was now putting my body through each day. I soon downloaded Strava to record my routes and measure my pace, and although I was now back at work and arriving home way after nightfall, the evening air was cold and electric, and I continued running every night. Moving along streets lit only by dull orange lamps, I could smell rain on tarmac, feel the icy wind on the back of my neck and the hot thud of my heart in my chest. When it wasn't raining I could still hear the *seep-seep* of redwings flying over my head, moving about the country in search of new territories. Winter would not last much longer, and soon they would be gathering together again in preparation for their long flight back to Scandinavia, where the ice would start to thaw and the earth would glow green with new life.

Britain would always be a warmer choice, and a few birds would remain in Scotland to breed, but our redwings will never stay. 'You don't understand, naturally,' says a swallow to the Water Rat in Kenneth Grahame's *The Wind in the Willows*:

> First, we feel it stirring within us, a sweet unrest; then back come the recollections one by one, like homing pigeons. They flutter through our dreams at night, they fly with us in our wheelings and circlings by day. We hunger to inquire of each other, to compare notes and assure ourselves that it was all really true, as one by one the scents and sounds and names of long-forgotten places come gradually back and beckon to us.

A quiet restlessness creeps into their hollow bones, and just as the swifts and swallows and cuckoos return to British shores each spring, our redwings will jump out over the sea and fly back to their Nordic homelands.

✶✶✶

The previous winter, Dave, his parents and I had flown to the western shores of India to escape the lethargy of Britain in late February. Each day the temperature climbed to 30°C, and we spent our time slicing through the streets on dusty mopeds, past stray dogs and sacred cows adorned with flowers, our senses overwhelmed by sunlight and spices. It was a beautiful and chaotic place, but each night, almost defeated by the noise and heat of the day, we would wander down to the shore and swim in the Arabian Sea where the cool, salty waters washed away the dirt and sweat. The air here was peaceful and still, but the waves pushed down on the shore in an endless rhythm, arriving and departing, embracing and withdrawing beneath the light of the moon. Under the surface I could feel shoals of tiny fish nibbling painlessly at my legs, reviving and renewing my body in the water. We tried to stand against the force of the waves, running to meet them as they swept up and crashed down onto the shore, but resistance was pointless. We were thrown into the water and carried back to the sand like driftwood. Trying to stand against the tide was like trying to stop time; better to float through it, uncontrolled, and embrace the rhythm of the water. We relaxed our bodies and the sea lifted us high into the air and back down to earth, ready to repeat that eternal cycle again and again and again.

One evening we rode to a small harbour hidden in a coconut grove and took a boat out to sea. We had been told there were humpback dolphins there, and as the sun started to sink below the horizon line we scanned the waves for any signs. I had never seen a humpback dolphin, so I was puzzled when our guide shouted and pointed to a small, grey triangle poking out of the water just metres from where we were floating. Unlike the dolphins of the British Isles (and all the cartoons) with their perfectly curved fins as long as their snouts, humpback dolphins have little stubs on their backs similar to those of the humpback whale. From the side they look a bit squashed, but their faces are still bottle-shaped and 'dolphin-like'. We watched one, two, three of them swim about in the water. These were *Sousa plumbea*, a subspecies of the Indo-Pacific humpback dolphin that are known to perform 'strand feeding', a communal feeding behaviour in which individuals work together to herd fish onto exposed sandbanks before deliberately beaching themselves to capture the doomed fish. Like other dolphins, they are also keen echolocators, using low-frequency whistles and high-frequency clicks to communicate through the water.

I was thinking about humpback dolphins one night in January, almost a year after our trip. Far from the warmth and spices of India, it had been snowing in Hampshire all day; a layer of muddy slush lined the pavements, and it was too cold to enjoy the outdoors. I was back at the flat having another cup of tea with Dave, and in between a third slice of toast and a 15th half-glance of affection we were both too afraid to fully commit to, I remembered the dolphins in the evening sun. All dolphins use echolocation to find prey, hunt, protect themselves and talk to each other in habitats that can often include murky, deep waters where their excellent eyesight is useless. They use the round part of their head – the melon – to send out clicks and whistles into the space around them; these sounds then bounce back off nearby objects, and the dolphins use special cavities in their jaws to interpret the echoes and build up a picture of their surroundings. It's a way to see through the darkness around them, and as I sat talking to Dave, listening to my own thoughts pulled from my mind and spoken out loud, I started to feel a new kind of clarity; a sense of what I wanted. Sometimes it's not until we vocalise our ideas that we truly start to hear them.

✶⁎

A few days later, two farmers were gathered around a tractor under the eaves of a large barn, surprisingly deep in conversation considering it was almost midnight. Dave and I had parked next to the Hare & Hounds pub in Stoughton, West Sussex, a pretty village with a Saxon church and a memorial to Polish pilot Bolesław Własnowolski who crashed his Hawker Hurricane there in 1940. From here we found our way along a track away from the road, moving into farmland and slowly towards the black cluster of trees standing over the horizon. Halfway along the track we reached the barn and the farmers, who looked upon our approach in a startled but not unwelcome manner and listened as we asked for directions.

'Kingley Vale? Up there!'

A hand pointed into the shadows and, before we could start pretending we knew the way, the tractor erupted in a blaze of light. The second farmer had kindly turned the headlights on to light the track and, waving a merry goodbye, we left the barn and proceeded up the long, sloping path towards the forest.

Kingley Vale is a nature reserve and Site of Special Scientific Interest on the edge of West Sussex, famed for its forest of yew trees that are among the oldest living things in Britain. Established in 1952, it was one of the country's first nature reserves and some of the yew trees that grow there are thought to be over 1,000 years old. These arboreal veterans were lucky to have survived, as some historians claim that most ancient yew trees in Europe were felled in the fifteenth century to meet the demand for a 'yew tax'. Yew wood was thought to be one of the hardest coniferous woods, perfect for the production of the English longbow that had grown to be a favoured weapon, so the government began to demand four pieces of yew for every cask of wine unloaded in an English harbour. It is unclear how much of this is fact or folklore, but large yew forests are still a rarity across Europe, growing in solitude or small clusters instead. Other species in the Vale include oak, ash, holly and hawthorn, and the chalkland beneath it all supports a vibrant mosaic of wildlife, including more than 50 species of birds, 39 different butterflies and lots of small mammals.

Dave and I had decided to explore Kingley Vale, to see if the rumours of ghosts and haunted barrows were true, and to enjoy some time alone away from the rest of the world. When we had first started seeing each other, one of our first afternoons together was a walk in the Downs during a rainstorm accompanied by his family Schnauzer, Tinks. Walking is the best catalyst for talking, and over the years during our walks we had laughed and cried, trudged through marshes, spotted butterflies, sipped coffee and listened to nuthatches *drip-drip-dripping* in the trees. When it rains, the countryside is almost deserted except for those stoic few who wrap themselves in wax jackets and persevere through the drizzle. The same goes for night walks, and we knew we'd have the black trees and trails of Kingley Vale almost to ourselves – the perfect excuse to start raking through the jumble of things in our heads and thinking about what might happen next.

The lights of the tractor guided us along the field and up to the foot of a long path that disappeared into the woods, and we started out into the darkness, leaving the glow of the village behind us. Tonight the sky was only half-clear, spread over with clouds so that the stars shone out between thick swathes of grey and the moon floated in and out of them all, lightening and darkening the landscape every few minutes. For a while we walked along the path, empty except for a number of gorse bushes whose silhouettes appeared suddenly from nowhere and frightened us, their sprawling branches like the limbs of crouching figures in the dark. Finally we reached the edge of the forest, a great, looming jungle, the scent of damp earth surging out from within it. We glanced farewell to the moonlit fields and crept forwards into the vault of trees.

Before we could even pretend to feel comfortable in the dark, a barn owl screeched to our right, the pearlescent banshee of the woods. Historically, barn owls were once Britain's most common owl species, but today many are killed on motorways or through eating rats that have ingested poison, and only one farm in around 75 is thought to host a barn owl nest. Conservation organisations are working hard to increase their numbers, aided by the fact that these owls are not territorial and will often overlap home ranges, meaning there is space for a lot more in our countryside. Their magnificent heart-shaped faces collect sound in the same way as the human ear, and ornithologists believe the barn owl's hearing is one of the most sensitive of any species ever tested.

To see one of these creatures floating through the sky, serene and pale in the half-light, it always comes as a slight surprise to hear them shriek out. Contrary to the myth that all owls hoot, barn owls emit a loud, high-pitched screech instead, giving them their common name: screech owl. In the folk tales of many European cultures barn owls were thought to be associated with evil spirits, and many were killed and nailed to the doors of houses afflicted with illness or death. Fortunately they are now a protected species, but I won't pretend it didn't give us a shock to be immediately yelled at by a fluffy white orb squatting in the trees above our heads.

The beauty of a forest as ancient as this one lies in the complexity of its layers, slowly woven together over thousands of years of growth to produce a unique and irreplaceable cocktail of living things. It's why it's so crucial to protect ancient woodlands and why, when they are destroyed, merely replanting new trees elsewhere is almost pointless. I remembered from my last daylight visit here how the ground was smothered in mosses, each one a differing shade of jade, turquoise, fern; springy and soft to touch, a vital part of the ecosystem. The Finnish author Tove Jansson wrote in *The Summer Book* of the frailty of moss, suggesting that it weakens drastically with every human footprint. There is truth in her words. Justin Bieber recently came under fire during filming for one of his videos in Iceland, when park rangers in Skaftárhreppur pointed out he had trampled over extremely sensitive moss, potentially causing irreparable damage. Some of the mosses in Iceland are so fragile that even footprints and tyre marks can take a long time to heal. Perhaps other species around the globe are a little hardier. They may not seem like the most exciting mounds of greenery, but mosses are crucial for maintaining healthy ecosystems, retaining water and humidity in their habitats, fixing nutrients in the soil and helping to control flooding.

Spread across trunks and rotting timbers, lichens are also one of the most underrated elements of the woodland, recognisable as those squidgy green things stuck to trees that are weirdly fun to peel off. At a glance they look like a plant or fungus, but the truth is they are the result of both of these organisms functioning as one, a symbiotic relationship formed between algae and fungi. Each one needs the other to survive; the algae produces carbohydrates through photosynthesis that feed the fungus, and the fungus protects and hydrates the algae. They can live almost anywhere in the world from tropical rainforests to the Arctic tundra, and they also indicate the health of the surrounding air, from which they absorb everything from carbon dioxide to heavy metals. Scientists consult lichens as the first indication of air pollution and toxicity, while garden birds peck beneath them to find insects hiding out of sight.

Beyond the mosses and lichens, larger plants colonise the ancient woodland floor, making use of the unique conditions that have evolved over their long existence on earth. Wood anemone and wild garlic love to grow in the shade, as does wood sorrel, dog's mercury and guelder rose shrubs, the latter producing succulent red berries that make an excellent homemade jelly. Between the plants and flowers, badgers, mice, stoats, foxes and rabbits weave their lives through the forest, while toads creep stickily through the leaf litter and bats sweep through the air in search of insects, eating up to 1,200 mosquito-sized invertebrates in one hour. And below their flight paths, holly and hawthorn spread out beneath towering native species like oak and (what remains of our) ash trees, and spindle bushes extend branches adorned with coral-coloured fruits that look like blushing nuggets of popcorn.

In the dark I could barely see the path in front of me, let alone a spindle tree, but I knew we had stepped into an ancient space. The square, predictable farm fields were left behind, and I could smell the energy, the dynamism of this place; roots and tubers buried deep in the soil, centuries of decay and regrowth, a sense of new life emerging from the old. We were lost in a labyrinth of shrub and thicket, our boots sinking into mud tracks as we walked, our ankles engulfed in a thick, writhing substance filled with living memory. How many creatures had walked, slithered, hopped and flown through these trees? How many people had fallen in love, murdered each other, laughed, cried and buried their treasure here? To wander into these places is to step away from our modern world; even the oldest towns and cities in England are infants compared to the oldest of our forests, where it's not only the trees that are veterans but the earth itself, feeding and sustaining the ecosystem with its collective of earthworms and bacteria. We think these are the lowest orders in the animal kingdom, but in reality they rule over us all; take them away, treat them badly and everything will be gone – the plants, the air, the water. It feels wonderful to protect the most charismatic members of the natural world, our giant pandas and clouded leopards, but in reality it's dirt that needs the most love.

The forest was so creepy I half-imagined we might be strangled by some devil-possessed ivy vines and dragged into the trees, a midnight feast for a gang of carnivorous plants lurking in the dark, but instead we just fell over. A lot. The mud was so thick that, without light, we slipped and toppled with every step, and when the path started heading uphill we could do nothing but laugh at our lack of dignity on the climb. Powerless in the dark, fear and frustration turned to silly joy, and we paused often to despair and eat squares of the chocolate I had packed in my bag. We swiped at the air, hands caked in soil, clinging to branches, to each other, squinting through fragments of moonlight to see if there was an end to the claggy trench that was taking us through the forest. All the while the barn owl lingered in the trees; every few minutes we heard a shriek in the canopy, and once or twice a tawny owl echoing from the top of the hill. The two sounds half-mocked, half-motivated us to keep climbing, and it worked. Finally, we emerged from the woodland and out onto the peak of the hill – and there, at the end of a small clearing, was the first of a black cluster of yew trees, limbs unfurled, needles shining, the source of so much mystery in this shadowy place.

Unlike Butser with its soft, smooth hilltop, Kingley Vale is still sprinkled with trees all along the ridge. Although there was now less cloud and more light filtering down from the sky, if we were going to encounter a Viking ghost anywhere, it would be here. It was light enough to reveal the outlines of trees and shrubs but dark enough to hide their identities, smudging them all into ominous shapes and threats to our primitive selves so that in an instant we had tightened up, made paranoid again by what enveloped us on the hill. And there, all around the grove, the yew trees crept and crawled in the darkness.

Of all the native trees in Britain, the yew is perhaps the most closely associated with 'sudden death', a phrase once used by the Ministry of Agriculture to describe the effects of digesting it. Almost every part of the yew is poisonous, and its poison known as taxine – used to kill Rex Fortescue in Agatha Christie's *A Pocket Full of Rye* – takes its title from the botanical name *Taxus baccata*. The only exception is the flesh of the fruit, which is edible to birds because the poisonous seeds inside pass through their digestive system untouched. In spring the female yew tree will bear flowers called arils, which contain one hard, toxic seed. As the season passes, these flowers encase the seed like a fleshy cone until they resemble a delicious red berry, when birds like waxwings and thrushes are enticed to eat them and help propagate the seeds elsewhere.

Gigantic and coniferous with reddish-purple bark, a yew can live for many thousands of years if left undisturbed;

some of the trees here were so old they had sunk their branches into the ground to create new roots, forming endless rings of yew trees so that parts of the grove had become one cryptic, tangled hollow. The reason for this change is because as the trunk ages it becomes more hollow, and the old branches sag into the floor where they can then take root. It is for this reason that the yew tree also represents immortality and strength, resilience in the face of adversity, and the ability to renew and regrow itself. In fact, there is no reason why an undisturbed yew tree could not continue this cycle of life and regrowth forever. It is a hardwood, durable, resilient and resistant to water, which is why in 1911 a spear made of yew wood was excavated in Essex that dated back an astonishing 450,000 years. It was also under a yew tree that the Magna Carta was signed by King John of England, 600 years ago in the water meadows of Runnymede.

The yew's reputation for holding the power of life and death continued on into early modern Britain, when Shakespeare included it as one of the Weird Sisters' cauldron ingredients in *Macbeth*:

> Double, double toil and trouble;
> Fire burn and cauldron bubble.
> ...
> Gall of goat, and slips of yew
> Silver'd in the moon's eclipse

Yew trees have also established themselves in Britain's old churches, the grounds of which contain some of our oldest specimens, protected among the gravestones. There is no one reason for this being the case, but historians believe some may have been traditionally planted on holy sites, while other theories suggest that, when Christianity took off in Britain, religious leaders thought it was easier to build new churches on old temple sites because people associated certain places with worship and would be more likely to transfer their beliefs over to the new religion. For this reason, many churches in Britain are built on burial mounds, tumuli and holy wells, and some historians claim that the yew trees growing in Christian graveyards may originate from an older time, when they were revered in pre-Christian belief systems. The ancient yew in the grounds of fourteenth-century St George's church in Crowhurst, Surrey, is twice as old as the church itself, and in 1820 the villagers held tea parties in its hollow trunk. A cannonball, thought to have been fired during the English Civil War, was found embedded within the trunk.

The clouds above Kingley Vale had almost vanished and the moon shone more brightly down onto the summit. Pale and glowing, the face of the moon can sometimes be used to count migratory birds travelling overnight, although it would take a patient eye to commit to a whole night of moon-gazing. The silhouettes of flying birds pass over the surface, and a skilled observer can count thrushes and geese passing over to new places, new territories. In the seventeenth century, the English minister and scientist Charles Morton wrote a treatise entitled *Birds in the Moon*, claiming that birds actually migrated to the moon and back every year, a trip he estimated would take 60 days if the poor birds could maintain a flight speed of 200kmph. He suggested that birds were not affected by gravity or air resistance and that they could complete the journey in two months by sleeping for 'most of the journey', sustained by excess fat. Watching the birds disappear each year into the endless sky, he could see no other solution to their absence than the idea that they must be leaving earth, asking: 'Whither should these creatures go, unless it were to the moon?'

Bird migration is often associated with movements of the sun, the warmth and cooling of the seasons, but many species do use the moon to shape their behavioural patterns.

A study of monogamous Barau's petrels on the volcanic Réunion Island in the Indian Ocean observed birds travelling to their mating sites over a period of time, and discovered they synchronised their journeys with the full moon, with the increase of moonlight triggering their hunting and mating instincts. In contrast, European storm petrels in Shetland use the cover of darkness to avoid being attacked by skuas and gulls, only returning to their nests in the dead of night. It's an amazing phenomenon, and one that is best enjoyed on the uninhabited island of Mousa, where a colossal Iron Age broch stands empty except for the thousands of storm petrels that swarm into the tower every night. The island is popular among summer visitors who come to see the petrels returning to their nests by the half-light of the midnight sun.

The broch itself is mentioned in two Norse sagas as a place of defence during invasions, as well as a lovers' hideout. By definition, a broch is a round tower with both an inner and outer drystone wall measuring around 5 metres thick in total, and this one – one of the best preserved of its kind – was built in Shetland around 400–200BC. Mousa flagstone is thought to be some of the best building stone in Shetland, and many similar brochs in the region were partially dismantled towards the end of the Iron Age, but miraculously this one remained untouched and archaeologists believe it probably stands just as tall today as it would have done 2,000 years ago. The *Orkneyinga Saga* tells how Erlend the Young once abducted the widowed mother of an Orkney Earl and held her hostage in Mousa broch; when the Earl tried to besiege the broch, he found it 'an unhandy place to attack'. Today, the island is designated as a Special Protection Area and managed as a nature reserve by the RSPB.

Although just a small bird, barely larger than a sparrow, the storm petrel has gained a reputation for itself in Western culture, particularly in maritime folklore. The word 'petrel' is a reference to Saint Peter, as the birds are so well adapted to ocean life that they can patter and dance over the waves, appearing almost to walk on water. The 'stormy' part of their name is thought to originate from their habit of hiding in the leeward side of ships to avoid the worst weather during storms. One superstition claims that storm petrels are bad omens, known as waterwitches, while in Brittany they are thought to be the spirits of captains who mistreated their crew, forever doomed to spend their days flying over the sea.

So strong an emblem is this bird that in 1901 it inspired a revolutionary poem by the Russian writer Maxim Gorky, 'The Song of the Stormy Petrel', or 'Pesnya o Burevestnike' in Russian. It was one of many allegorical fables written at the time to declare support for revolutionary ideas without openly speaking out against the Tsar. In his poem, Gorky writes of a proud 'stormy petrel', unafraid of the turbulent storm in which it flies – the revolution – while all other birds cower in fear, as translated by Eugene M. Kayden:

The stormy petrel
Soars between
The sky and deep.
He rides the waves,
And like a flash,
A thunderbolt,
He strikes the cloud-ranks.
Joy defiant
Hear the clouds
In the petrel's crying:
Thirst of tempest,
Flame of anger,
Might of passion,
Faith of triumph,

Hear the clouds
In the petrel's crying!
…
Alone, sublime,
The stormy petrel
Soars free above
Wind-cloven waves
Lashed white with anger.
Lower, blacker,
Hang the storm-clouds
On the ocean;
Higher dance
The waves in frenzy,
And leap to meet
The blast of thunder.
…
Alone the proud,
The stormy petrel
Over the spouting
Savage sea,
Alone he soars
A prophet crying
Of victory:
Let the storm rage!
Fiercer,
Let the storm break!

A powerful piece, although sadly Gorky's symbolism wasn't quite deceptive enough; he was arrested during the short-lived 1905 Revolution, and spent a large part of his adult life in Italy, exiled from Russia and the Soviet Union until he was invited back by Joseph Stalin himself. Nevertheless, this poem captures the enchanting nature of such a small and largely unnoticed seabird, and it is no surprise that the midnight boat trip from Shetland to Mousa is fully booked every year.

Huddled in the trees in the heart of Sussex, we were a long way from the storm petrels nesting in Mousa broch, but the same Viking spirits that once inhabited Shetland were still among us nonetheless. Nobody knows for certain why Kingley Vale has become so closely associated with Vikings, but plenty of theories have been put forward. According to the ancient *Anglo-Saxon Chronicle* manuscript, the Vikings made their first raids on English shores towards the end of the eighth century, when they attacked both the Isle of Portland in southern England, and Lindisfarne in the north-east where they killed all the monks and looted their treasures. Their role in British society changed as the decades passed: at times they were executed as heathens, at others welcomed for their tradeable goods, and at one point they were even paid as hired thugs to protect the rest of the country. Nevertheless, their time in Britain, like most migratory shifts throughout history, came to an end with the rise of new powers and authorities, but they left their mark on town names, customs, festivals and beliefs, as well as the genetic make-up of the British people themselves. My mum has traced our family tree back to the Orkney Isles so I'm hoping we have Viking blood in our veins.

The most popular local Viking myth is connected to the Devil's Humps nearby, four Bronze Age barrows at the top of the Vale. Historians believe the men of Chichester once defeated a Viking war party who had taken hold of the Vale, and when all the leaders were slain they were buried beneath the Humps. (However, as the Humps have been dated back to the Bronze Age, this might be a slight historical smudge.) The rest of the Viking dead were left unburied where they had fallen under the yew trees on the slopes of the hill, and their ghosts are said to still haunt the yew groves to this day while the trees come alive and walk through the forest at night.

It became easy to imagine other people lurking ahead of us, every twig and shrub contorted into pale hands and hidden faces, every shadow enlarged in the brightening moonlight. Not only does the night conceal, it transforms things we once recognised by daylight, mutates and changes them into something unknown, unwelcome. We spotted a trig point in the distance beneath a crown of trees and stood, motionless, our minds alive with fear until we realised it was just an innocent little beacon asleep under the stars. The night frightened us in Kingley Vale, but there was something constant about it, too. Something reliable and reassuring about night always following day, something to be not conquered but admired. By the end of our walk the fear had subsided, the paranoia faded, and we were almost relaxed in the darkness, unafraid of what it was shielding.

There's nothing like an eerie walk in the dark to bring two people together, and by the time we emerged from the forest and back onto the track, we had returned to our past selves. Being unable to see each other's faces, we instead relied on changes in tone, movements masked by darkness, those natural, uninhibited behaviours that cannot be concealed by cleverly worded sentences or the urge to 'play it cool' – no matter how hard you try. Awkwardness faded to peace, and with it joy came swimming back to us, the sense of being around someone who makes you fall so in love with the world that all the loneliness of being alive on earth transforms into something new; two souls joined together with all the other atoms in the universe; an infinite, effervescent reverberation between us and the stars.

We wandered back down the Vale, leaving behind the yew trees and emerging into clearer land, scattered with young hawthorn trees, spreading out into farmland once again. I have always loved living in Hampshire, but there is something about Sussex that's a little more charming; more wildflowers in the hedgerow, birds in the trees; fresh eggs for sale on pastel-blue doorsteps; quiet rivers and drystone walls.

Sussex seems to be allowed to grow wilder. The village of Stoughton lay sleeping at the end of the track, a handful of lights fighting back the shadow; as we neared the road, our barn owl gave one last shriek, far beyond the hill, before the dark forest disappeared from view. We had survived the mud and the Viking ghosts, and hand in hand in the moonlight, we were finally going home.

Royal
Observatory
Greenwich

CHAPTER FIVE

Greenwich

One beautiful morning in late January, I was walking through Greenwich Park on the south bank of the Thames, around that area of the Underground map where everything sounds pirate-based. Cutty Sark, Crossharbour, Pontoon Dock; the history of London is already enticing, but this place was so marine-themed it seemed to jump right out of a Joseph Conrad novel. In fact, one of Conrad's later stories *The Secret Agent* takes place in the park itself, when a spy called Adolf Verloc is ordered to detonate a bomb and blow up the Greenwich Observatory – an attack on modern culture and a reminder of the dangers of underestimating anarchism.

The novel is based on a real explosion that happened in 1894, when French anarchist Martial Bourdin accidentally blew himself up when his bomb detonated too early. His motives have never been clear, but historians suggest he was trying to destroy the 24-hour gate clock that had been displaying Greenwich Mean Time to the public for over 40 years, a symbol of rebellion against government control. Once staff had finished clearing away the bits of flesh and bone, the Observatory's chief assistant Herbert Hall wrote a simple note in his journal: '16.45: A dynamiter anarchist was blown up with his own bomb in Greenwich Park.'

Almost 180 years later, the Royal Observatory is still standing, a beacon of modern astronomy and navigation now open to the public all year round, and I had come to Greenwich to see it. Over the centuries we have mapped out the stars, worked out the age of planets and even travelled into space, and for a long time I had wanted to visit one of the places where it all began. It was no longer the official working observatory for British astronomy, as high light-pollution levels had resulted in the decision to move to Herstmonceux Castle in East Sussex after the Second World War, but it was still an emblem of science and heritage, and I was excited to see the antique telescopes and navigation devices that had laid the groundwork for modern astronomy.

The temperature that morning was uncomfortably low, but the sun shone like gold on the grass managed so neatly by the park authorities, and in the trees I could see the earliest buds beginning to emerge, the start of new life and a new year. One hazel tree was already speckled with catkins, the name given to the male flowers of the tree that look like dangling, olive-coloured caterpillars.

The Observatory was hidden away up a steep hill and the incline warmed me up as I walked. The building itself was developed from the ruins of Greenwich Castle, built in the 1430s and reportedly the favourite place for Henry VIII to house his mistresses as he could travel easily to see them from his other residences. As the Castle foundations had still been strong, the architect and astronomer Sir Christopher Wren suggested using the ruins as a site for the new observatory, which had been commissioned in 1675 by King Charles II and his board of royal advisers. Greenwich remained the official location of the Observatory until 1924, when some of the departments moved away to other parts of the country. The reason for this was that the electrification of the railways that year had caused some of the magnetic and meteorological readings to go askew, and in the time leading up to the relocation of the Magnetic Observatory to Surrey, they even had to insist that electric trams in the surrounding area were forbidden from using an earth return for the traction current.

When the Observatory was first built, Charles appointed the first Astronomer Royal, John Flamsteed, to start plotting all the stars visible in the northern and southern hemispheres. At this point, large portions of the rest of the world were controlled by the European empires, and although trading with other countries was growing more and more successful, the astronomical information needed to navigate the seas was lacking. While captains could easily measure the latitude of their location at sea, it was almost impossible to measure their longitude once out of sight of land, so in 1714 Parliament passed the Longitude Act, offering large rewards for anyone who could find a practical method of determining a ship's longitude.

Around 50 years later, Nevil Maskelyne, the first person to scientifically measure the weight of the planet, made a breakthrough. He oversaw the publication of the *Nautical Almanac*, a set of carefully observed tables that finally enabled sailors to determine their longitude at sea. He recognised how the moon, a fixed point in the sky, could be used as a consistent reference point for ships – providing the sky was clear – and went on to devise a calculation known as the lunar distance method. The lunar distance is the angular distance between the moon and another celestial body (like a star), and by comparing this angle with the local times recorded in his *Nautical Almanac*, navigators were finally able to approximate their position. After Maskelyne's initial success, the famous clockmaker John Harrison went on to design and create the first accurate marine timekeepers, portable clocks that kept time to within three seconds of accuracy.

As astronomy and time are so inextricably linked, the data recorded at the Observatory was vital in developing accurate timekeeping on land as well as at sea. By the late nineteenth century Britain officially declared that all clocks in the country would be set to the Greenwich Meridian rather than the array of local times kept by different towns, and in 1884 the Greenwich Meridian was chosen to be the Prime Meridian against which all the world's clocks would be set. To this day, the Royal Observatory Time Ball signals to the watching public the exact time every day at 13.00. At 12.55 an ancient, orange ball is hoisted over London, ready to be dropped at exactly the right time, which would have once allowed sea captains waiting on the river to check the rate of their marine timekeepers before sailing out into the ocean. I later discovered you can buy a jar of your own aniseed 'time balls' in the gift shop on your way out.

Walking through this postcard-perfect observatory in Greenwich Park, I couldn't stop my mind wandering down to the layers of history beneath my feet. A few metres away on the bank of the Thames, London's history is literally dragged out by the shifting tides, and people have scampered about for hundreds of years in search of valuable items exposed by the mud. The hobby is called mudlarking, and was particularly common during the eighteenth and nineteenth centuries, when poverty and lack of education enticed people of all ages to make an income from scouring the banks at low tide. From padlocks, coins and musket balls to pottery, fossilised shark teeth and Mesolithic axe heads, freshwater shores can theoretically produce anything that has passed through the waterways – and for the River Thames, that could mean anything.

As for inland treasures, the roads in London have been built and rebuilt over 2,000 years, a multi-layered sandwich of concrete, stone, earth and every dropped possession that's slipped in between. Although the Romans are credited for inventing the road system, building 87,000km of road throughout their empire, many archaeologists believe they probably stole the idea from the people of Carthage before they burned the place down and sowed the land with salt in 146BC. Before the Romans arrived in Britain there were no traditional roads in place, just the green tracks and dirt paths that still exist today. Most roads were designed to connect military camps, and since their main purpose was to carry foot soldiers long distances, they were straight.

But it wasn't just roads we have the Romans to thank for. Around 100AD, Claudius Ptolemy was born in Egypt as a

Roman citizen. He would later become one of the greatest scholars of his time, making enormous contributions to the fields of astronomy, mathematics and geography, but he is now remembered just as fondly for his wrong theories as his right ones. He famously claimed that our earth stood still in the solar system, and that the sun, stars and moon all orbited around it. Although some earlier astronomers like Aristarchus had already started to theorise that the earth travelled around the sun, Ptolemy wasn't convinced, and his consequent theories about how other planets moved were all wrong because his calculations were based on the earth being the centre of the universe.

However, astrometry – the most ancient branch of astronomy – started thousands of years ago far from the future stronghold of Europe, primarily in China, Mesopotamia, Central America and India. Astrometry was the art of measuring the sun, moon and planets, the precise calculations of which allowed observers to model the evolution of planets and stars and to predict phenomena like meteor showers and comets. These early astronomers noticed patterns in the sky, and tried to record their movements and understand their order. The first patterns to be recorded are now known as constellations, the arrangement of stars that helped our ancestors measure the seasons, giving birth to the hundreds of ancient stories surrounding the night sky today.

NASA defines astronomy as 'the study of stars, planets and space'. Modern astronomers are divided into the observational and the theoretical: some focus on observations of physical bodies like planets and stars, and some use measurements to analyse systems and try to understand how the universe evolved. Unlike so many other fields of science, no astronomer (or human) is able to observe a galactic system from birth to death, as even the tiniest star will exist for millions of years at least, so we have to rely on snapshots of the universe in various stages of evolution to work out how they formed and when they will die.

Of the many branches of astronomy, cosmology is one of the newest, coined in 1656 by the English lexicographer Thomas Blount. It focuses on the universe in its entirety rather than individual objects, theorising its existence from the Big Bang to the present, and then onwards to its inevitable fate. Whether or not we will be there at the end of the universe is a mystery, and one that perhaps we should not solve. The knowledge we've gained as a species is both a gift and a curse, and it becomes easy to envy other creatures who are not burdened with the inevitability of what might happen to the universe, but who spend their days looking for food or sex or a warm place to sleep. They don't worry about powers out of their control, nor do they philosophise if nature is cruel or kind, an unstoppable force that reacts to unpredictable events and could cast an entire planet into oblivion without hesitation.

I watched *Interstellar* for the second time the other day, a Christopher Nolan film that follows a group of astronauts as they travel through a wormhole in search of a new home for humanity. It struck me how desperate we are to colonise other planets rather than focusing on the one we have. Not that I'm against space travel – if somebody offered me the chance to go into space, I would snatch it. I remember visiting the Kennedy Space Center in Florida when I was five, and watching a shuttle leap away from the earth and into the great unknown. How amazing would it be to leave the planet and float through a silent galaxy with nothing but stars and space dust for company? To know that beyond the walls of the little metal cabin you're sitting in, there is nothing but an empty void, a starlit chasm? To bob around without gravity and eat freeze-dried strawberries? Yes please. But it's worth remembering that in 2014, NASA had a total budget of $17.6 billion, the 2012 US defence budget was a jaw-dropping $737 billion, while in 2008 the United Nations reported that just $30 billion a year could end world hunger.

I finished my hike up to the Observatory, crossed the Meridian Line marked on the floor and wandered into the Astronomy Centre. Here you could touch a 4.5 billion-year-old meteorite (as old as the sun!), and I clung to it with relish. I love anything ancient; an archaeologist friend once let me hold a real mammoth tusk and I couldn't handle the excitement of touching something so real, an actual fragment of a once-living, hairy, pungent, beautiful mammoth. This meteorite was grey and smoothed by the millions of human hands that had run across its surface. Beyond this the building was divided into dark, cosy rooms filled with screens and a range of ancient timekeepers and navigational devices. Brass spectroscopes and sextants shone behind panes of glass, relics of an older world when their invention would have been worthy of a slot on *Tomorrow's World*. I imagined these devices being displayed to society hundreds of years ago when the city was smaller; will this be how our great-grandchildren feel when they see the first satnavs and Bluetooth headsets behind glass in some ancient technology museum?

'How *clunky* it all was, Grandmother.'

'You actually had to *ask* it to avoid toll booths!'

'What's a U-turn?'

With its design-led modern skyline and an unofficial bid to become Europe's 'silicon capital', London has become a beacon of progress and technology. But beneath it all lies a hidden city, an older world sleeping under the buses and shops and skyscrapers.

The city of London was founded in 43AD when the Romans established a settlement and major port called Londinium on the banks of the Thames. Early London was tiny, around the same size as Hyde Park and a far cry from the 1,569 square kilometres that now makes up the region of Greater London. In its early days, the Iceni queen Boudicca's rebellion forced the Romans to abandon their new settlement and it was razed to the ground, only to be rebuilt when they defeated the Iceni at the Battle of Watling Street around 61 AD. By the end of the first century it had expanded rapidly, eventually becoming the capital of the United Kingdom, a global leader in the arts, entertainment, fashion, healthcare, media, research and tourism, as well as the largest financial centre in the world. Those Romans would have been proud.

When I was studying for my Masters degree I lived in south London with friends, but to escape the claustrophobia of the city I spent every Thursday volunteering with the London Wildlife Trust at one of their reserves, Sydenham Hill Wood. It was once part of the extensive Dulwich Woods and the largest remaining tract of the ancient Great North Wood, a natural oak woodland that stretched from Camberwell almost down to Croydon. The Great North Wood was lost to urbanisation over the centuries, and in 1854 Sydenham Hill Wood was dissected by a steel railway carrying passengers to the newly built Crystal Palace, a cast-iron and glass structure originally built in Hyde Park three years earlier. The Palace was designed to house the Great Exhibition, an international world fair to display London's hoarded treasures from around the world, including the famous Koh-i-noor diamond and the first fax machine. A similar fair in Paris in 1889, called the Exposition Universelle, is still famous for the specially designed entrance gate that was left there after the fair closed, now known as the Eiffel Tower.

After three years the London Exhibition grew in popularity until it was moved to Penge Peak in south London, where it stayed until 1936 when it was destroyed by a fire. There were so many visitors to the Palace that a new railway had to be built through the wood to carry the crowds, and the area became so popular that a number of fashionable villas were built along the edge of the trees, a hive of social merriment. According to one story, an elephant being housed in the Palace once escaped into the wood and rampaged across the footbridge before being recaptured. When the Palace burned down, the railway closed and the houses became derelict, and over the decades nature has slowly been allowed to reclaim her territory with help from the London Wildlife Trust.

When I started volunteering there, the woods were still littered with relics of fashionable gardens, rockeries and sundials hidden behind wild birch trees. A folly was once built to simulate the ruins of a monastery, and a pleasure pool was carved into the earth like a bomb crater. There were also rhododendrons that once filled the borders of flawless lawns, a Chilean monkey puzzle tree and a cedar of Lebanon left to rise alone above the canopy. Now the wood was wilder, overgrown, but there were glimpses of the past growing under the gaze of firecrests and tawny owls. Even the railway didn't fully disappear; we walked over the footbridge and a nettle-strewn valley was carved out of the ground beneath it, the occasional rusted train track lying forgotten, corroding beneath the earth.

Volunteering in the wood was the best part of what became an isolating seven months in London. The city is old and alive, a gigantic machine full of people and music and stories, and I enjoyed travelling into the centre each week to study great writers and poets, trying and yet again failing to read *Ulysses*, discussing the intimate details of every paragraph in the way that only English-literature students can. But then the day was over, and I had to retreat back to the outskirts, away from the vibrant core of the city. London has a beauty of its own, but I recognised that I couldn't live there. I needed the chalklands of the South Downs, the beech trees and birdsong of the countryside where nothing much happens, the bustling simplicity of a small town where the most exciting news of the decade was when Waitrose started giving out free coffees. The city of London glitters; it's full of wonder and movement like a waterfall – loud, beautiful, constant. But for me, it is not raw beauty like the roaring coast or the forests, where the air swells with life and nature weaves itself into an infinite tapestry of growth and decay. To watch dandelions creep through pavement cracks and peregrines tumble from the Tate is wonderful, but to stand in a meadow and feel bees brush against the hairs on your arms, to smell wild roses and listen to the skylark ascending into the ether – that is sublime.

It was in winter that I spent most of my time in Sydenham Hill Wood, when the leafless trees were tangled boughs in the dark. Along the old train track the ground was undisturbed and carpeted with fungi: candlesnuff, oysterling, birch polypore, coral spot and clouded funnel, and electric-blue spores of cobalt crust grew unnoticed on fence posts. In winter the nights came early and, within the shadows of the trees, there was a sense of mystery about the place, as if the forest was a secret that the rest of London hadn't heard about, and it was up to us to keep it safe, protected, alive. The earth was soft and fragrant underfoot, the air full of insects.

One evening our volunteer group arrived later than usual to complete a bat survey. As darkness cloaked the trees, we heard the last diurnal creatures moving around us; green woodpeckers undulating between branches, a nuthatch on the mulberry tree, wood mice creeping silently over mouldering leaf litter. The purpose of the survey was to keep an eye on the bats that were roosting in the abandoned train tunnel now closed off to the public. We were given high-vis jackets and torches, and then made our way over to the tunnel entrance, near the shipping container where vital tea and biscuit supplies were kept. We were accompanied by a licensed bat expert, and the intention was to find the bats roosting in the tunnel, record a few observations and then leave them in peace to enjoy their hibernation. It was late winter and bitterly cold, and being one of only three mammals to hibernate in Britain – along with dormice and hedgehogs – it was the perfect time to sneak up and count them without causing too much disturbance. The results would be fed back to the London Wildlife Trust and the Bat Conservation Trust to help them gauge the health of London's bats and work out how best to preserve and protect their populations.

Bats have a hard time in the human world. One summer I volunteered for the Bat Conservation Trust's bat helpline, answering calls from people who had a bat in their house, caught by the cat or from a roost in the roof. It became clear that, although most callers wanted to help wildlife, they had more difficulty coping with a bat than a bird or rodent. Many of us grow up believing bats carry rabies and can get tangled in your hair (too many reruns of *Ace Ventura*), but the truth is they are excellent fliers and would never be stupid enough to fly into your head, let alone somehow get caught up in your hair. The rabies myth is not entirely unfounded; each year a few bats in the UK are tested positive for a rabies virus called European Bat Lyssavirus, but this is not the classic rabies strain. It can be treated easily with vaccinations or medical attention, but treatment is only necessary if you are bitten by the bat, which is unlikely unless you're a licensed handler.

Perhaps it's because they are nocturnal creatures that they have gained an unsavoury reputation, being associated with blood-sucking and vampires, but British bats are incredibly sweet on closer inspection, being essentially no more than velvet mice with wings (although they are more closely related to humans than mice). They are incredible creatures, perfectly adapted to the night through their use of echolocation to navigate and hunt for insects in the dark;

even the smallest bat can eat up to 3,000 insects a night, so they are a welcome form of garden pest control. In tropical countries they also feed on fruit and flowers, and it's thanks to bat pollination that we can enjoy dates, vanilla, bananas, guava, tequila and chewing gum. Unlike other small mammals, they will only have one baby or 'pup' a year, which means it's even more important not to disturb their roosts, as a healthy bat can live for 30 years if left in peace. Aside from a small handful of predators in the wild, one of their biggest living threats is the domestic cat, which is known to catch individuals and kill, eat or maim them, just as it does with our dwindling garden bird populations.

At the entrance to the tunnel somebody had written 'Moria' in graffiti paint, which I thought amusing. When we climbed through the doors and into the tunnel we found newspapers lining the edges dating back 50 or 60 years to when the line had been closed. Shut off to the public for both safety and conservation reasons, it was amazing to be allowed inside this time capsule of an older London. As we wandered through, our torch lights were reflected off broken timbers and glass that must have been lying there for decades.

The survey technique was simple and efficient: we spread out and methodically made our way down the tunnel, checking every tiny nook and crack in the wall that might be home to a sleeping bat. Despite their wingspan, bats are tiny creatures and will fit into the smallest gap in order to feel safe and hidden from predators. Many species like underground sites like these – known as hibernacula – because they are less likely to be disturbed by light and noise, and will often provide the optimum humidity and consistently low temperature the animals need to survive a winter hibernation.

When one of us found a bat, our expert would hurry along and carry out a few observations before moving on to the next one. We walked slowly, and it was extremely cold but exciting to be exploring such a dark and peaceful sanctuary for wildlife. At one point I even found a butterfly hibernating on the wall, tucked in a corner, wings closed, dreaming of wildflower nectar and sunshine. We finished up in the tunnel and returned to the world above ground, safe in the knowledge that sleeping away in a corner of an ancient wood in urban London, a small population of bats was alive and healthy, ready to wake up for spring in a few weeks' time.

One of the things I found most difficult about living in London was its sleeplessness. No matter the hour or the place, there is always movement on the air, sirens in the background, trains running, buses stopping, so many people walking through the city on their way to somewhere new. In lots of ways this is a good thing – since leaving I've never again been able to get pizza at 3am – but after a few months I started to find it draining. I missed the peace of night, the space to rest and recover. It wasn't that I suffered from insomnia or that the neighbours kept us awake, but when I went to bed I knew the city was still alive, never ceasing to work, like an endless, incessant machine. Back home the streets would be quiet; birds asleep in their nests, the trees cloaked in darkness, stars pouring their energy out into the night. Despite my love of being awake at night, we are diurnal creatures, and I love the pause between one day and the next, one brief moment of peace, of reflection. London was exhilarating, but by the time I left I was exhausted by it, desperate to return to a place where I could see the sunrise and hear the dawn chorus at its natural time, not drowned out by exhaust fumes and club music.

I thrive on peace. I love adventures, travel, nights out, all the things I should love as a twenty-something, but I also need peace and solitude. One of my favourite places to wander is Old Winchester Hill in the South Downs, a hilltop forest and Bronze Age burial site that looks over the rest of the landscape, where you can watch kestrels and kites floating through the air currents in the valley below you. Connection with nature means different things to different people, but for me it is meditative, a space to think and feel what I'm naturally inclined to think and feel, to absorb the energy of my environment. When I was in London, I found the energy stimulating but overwhelming – too much life and chaos with very little time to reflect on my day-to-day life and what I wanted to do in the future. Now I'm back home in the countryside, one of my favourite parts of a night in town is the walk back from the pub on a warm evening, past an old brick house with honeysuckle growing on the wall outside. In the balmy air of the night, the aroma of those flowers engulfs my senses and carries me all the way back home in peace.

And yet London is one of the greenest capital cities in Europe, with one estimate claiming around 47 per cent of the city is made up of green space. From hedgehogs in Regent's Park to almost 200 species of bird recorded on Hampstead Heath, from peregrine falcons on the Tate to the more exotic rose-ringed parakeets squawking in park woodlands – between the bricks and concrete, wildlife is still thriving. One of my favourite species, and one that is both loved and hated across the country, is the fox. According to fossil evidence, the modern red fox has been living in Europe for at least 400,000 years, and is one of the world's species that appears to have evolved as a result of convergent evolution. This is where two species look extremely similar, not because they are from the same genetic family, but because they occupy the same ecological niche or habitat and have developed similar adaptations. One example includes dolphins and sharks, which look similar but are not related, and the same is sometimes said with foxes and cats. Although foxes are from the same family as dogs, observe them for a period of time and you will notice they have the same delicate gait as cats, the same habit of stalking and pouncing, the same curled-up position when asleep, and the same twitching tail movements.

I remember seeing a scrappy dog fox on the way home from a night out once. We'd been at a music night in Brixton, the weather was mild and so, rather than spending money on the bus back to our flat in Streatham, we decided to grab a bag of hot chips from the kebab shop and walk home. The sky was clear after a long, warm day, but I could only see one star amid the glow of the city lights, right on the edge of the horizon so that it was barely there at all. This was Sirius, the brightest star in the sky, also known as the Dog Star as it is the brightest in the constellation Canis Major. Aside from Sirius, the night sky was stained with that rusty glow that bursts out of the heart of the city and into the air, diluting the darkness and shielding the stars from the naked eye.

We had been walking for a few minutes, past the main buzz of Brixton and down to a quieter part of the road, when a dog fox leapt over a brick wall a few metres ahead of us. A few cars were still rolling past, and he stood frozen on the wall, watching the traffic, considering where to move next, with both his ears and tail – surprisingly bushy considering he was a city-dwelling fox – rigid and alert. Still scoffing chips and feeling relaxed by whatever was currently coursing through our bloodstream, we watched and waited to see what he might do, and after a few seconds passed he was off, scampering across the road between the cars, disappearing into a dark side alley with a flash of his tail.

I loved seeing this fox on our way home, a glimpse of wilderness among the grey lights of south London. A few weeks later I was taking the rubbish out behind our block of flats when I found something even better. A whole family of foxes was living on our estate, lingering suspiciously by the bin shed, scrapping outside the fire exit, yapping at night when most of us were asleep. There were three adults and a couple of cubs living outside our flat, and when I first went to take the rubbish out they scurried away, wary of humans but reliant on their waste for nourishment.

Urban foxes are thought to be more prone to mange or fleas, presumably because they feed on our waste rather than on more natural prey found in rural areas, which makes them more likely to pick up diseases. This means they can look more dishevelled than the fluffy foxes of the countryside, but the truth about mange is that it can be easily treated with medication. It's a disease caused by a mite called *Sarcoptes scabiei*, and when left untreated it will eventually lead to the death of the fox. But while a mange-infected fox can look emaciated and starving, these symptoms are just caused by the fact that the fox is so irritated by the infection that he can no longer look after himself, which eventually leads to further weakening and death. Some wildlife charities provide treatment for mange free of charge, which should be placed in a batch of jam or honey sandwiches in the garden for the fox to eat.

Fortunately the foxes outside our flat looked healthy and plump, and each night I started to look out for them from my window. One night I found them playing on the ground directly below our bedroom window, so I decided to try and feed them. We didn't have any specific fox cuisine, so instead I sprinkled peanuts from the window and watched what they might do. Alarmed at first by a raincloud of nuts, they soon grew curious and wandered over to the dropped goods, eating them one at a time off the ground. The next day I walked down to the pet shop and bought a bag of dry dog food, and soon I was feeding the whole pack of five outside – one or two more reserved than the others, and the alpha animals gathering most of the food for themselves.

Feeding the foxes brought me an enormous amount of joy, especially as by this point I had started to realise London wasn't making me happy. Being without much money,

I found it more and more isolating by the day and, although my friends and boyfriend were great to hang around with, I hated not being able to leave the flat without spending money on transport. I walked down to Streatham Common from time to time but, while it was pretty and green, I couldn't overlook the fact that I was surrounded by the city, closed in on all sides. Perhaps I had been spoilt by growing up in such a rural area, but it meant that a city park was hardly a substitute at all; it was like going to an indoor rock-climbing centre instead of climbing a real cliff face outside in the wind and sun. The Common provided me with wildlife and fresh air, but I still felt enclosed by the city and I was always aware of what lay beyond the park's boundary.

Staying in Streatham for most of the week, the foxes brought a touch of the wild to my own back door. I wanted to try and get closer to them, to see how wary of humans they truly were. I didn't want to tame them – even if I could – because I knew my place there was only temporary, and I didn't want to put them in danger of interacting with other people who might try and hurt them. But I wanted to see how close I could get, at least on ground level rather than peering down from my bedroom window.

The following day I was walking back from the bus stop and decided to see if I could find something more tempting for the foxes to eat. I found a packet of lamb's liver in the corner shop for 75p, and that night I decided to walk down to the fire exit and wait quietly for the foxes to appear. It was early autumn and still warm, so I sat and waited, voices floating out of the windows of other apartments. After half an hour, a fox appeared from behind the bin. It was one of the teenage cubs, probably born that spring but now almost fully grown, and I could see he was curious with that invincibility of youth. I took a lump of liver from the packet and gently threw it over, which made him run away again behind the bin shed. I waited. A few minutes passed, and a head reappeared behind the wall, staring at me and the lump of meat on the floor.

One paw emerged from behind the wall and stepped forwards, followed by another, more hesitant, and slowly the fox walked back over to the piece of meat and stared at it, head cocked, ears twitching. As I watched, he bent his head down, sniffed the air and started to eat the slice of liver. I sat motionless, hardly daring to move but aware that he would finish it and be off again within half a minute. He looked over to where I was sitting, but just as I reached for another piece, he ran away. I threw the other piece over and, again, after a few minutes he returned to the same spot to eat it up. Again I threw a piece of meat and again he retreated, but this time not all the way behind the wall, just far enough to gauge the situation. And this time, when he returned to the same spot, he was followed by one of the other foxes, a larger one this time – a grown adult.

I spent the night feeding them slivers of offal, and by the time it was all gone my hands were soaked in fresh blood, the foxes' bellies full of the scraps of meat nobody else wanted. Despite my concrete, treeless surroundings, and despite the fact that the light pollution had washed away the stars, I was full of joy at being able to connect with something so wild in the heart of the city. These foxes were ragged and mischievous, and I was probably breaking some city law by feeding them, but for the first time in a while I was able to feel the same euphoria that comes from enveloping yourself in nature, the feeling I had every day at home that I had been sorely missing since moving to the bright and busy city.

Unsurprisingly, I didn't last much longer in London after that. In January I spent a weekend at home for my birthday and realised how miserable I felt going back to the city, so I broke up with my boyfriend, moved out of the flat and back in with my parents. Not exactly every twenty-something's dream, but I was so happy to be back with my family, back in the countryside, back in our silly market town where too many people read the *Daily Mail* and everyone cares way too much about bin-collection dates. This was where I wanted to be, and although it was strange to be back to the start, back to my hometown where I'd kissed my first boyfriend and drunk my first bottle of vodka, it felt right. Then, around a year after moving home, I met Dave on a night out for a friend's birthday, and everything fell into place.

TIF

CHAPTER SIX

Under Dark Skies

Easter has never been a biblical time for me; I have an interest in religious stories from a cultural point of view, but I get more excitement from chocolate eggs, daffodils and simnel cake. The best part about Easter, however, is the four-day bank-holiday weekend, and one year – always keen to make use of extra time off – we slipped away to the West Country to explore Cheddar Gorge, the Somerset Levels and Exmoor.

We spent the first day wandering through Cheddar, watching feral goats clambering up the cliff edges, feeling the cool dampness of the rocks that descended underground and formed the famous Cheddar Caves. I remembered coming here as a child, when we visited a series of limestone caves known as Wookey Hole, formed over thousands of years as the natural acid found in groundwater had slowly dissolved the rocks. Archaeologists believed they had been used by humans for around 45,000 years, with Palaeolithic tools found among fossilised animal remains, as well as evidence of Bronze Age, Iron Age and Roman occupation. The caves have a constant temperature of 11°C, creating the perfect coolness and humidity for maturing the now-famous Cheddar cheese.

One of the stalagmites that had formed inside the caves was now known as the Witch of Wookey Hole, originating from an old legend about a monk from the nearby New Age town of Glastonbury. The story goes that a witch once lived in the darkness of Wookey Hole and, jilted by her lover, had vowed to spend the rest of her days putting curses on young couples in love to sabotage their happiness. One day, a man became engaged to a girl he loved in the village of Wookey, but not long afterwards the witch found them and cursed them, causing their romance to fail. The man believed he could never love again and became a monk, while seeking revenge on the witch who had ruined his happiness. He stalked the witch into her cave, where she hid in the darkness by an underground river. The monk blessed the water and splashed it into the shadows, where it fell upon the witch and immediately petrified her into stone, and she remains there to this day. The story is also connected with the 1,000-year-old skeleton of a woman discovered in the caves in 1912. The remains were excavated and now lie in the Wells and Mendip Museum, although some believe her spirit will not rest until she is brought back to the caves where she belongs.

After a long day climbing through Cheddar Gorge, on Easter Sunday we woke up next to the King's Sedgemoor Drain, where the night before we had parked the camper van to watch the sunset over the water with two bottles of the local cider. We took an early-morning swim in the river, and then walked up to see where local conservation groups had installed eel passes on one of the sluice gates. The Drain is an artificial drainage channel that diverts water from the peat moors of King's Sedgemoor to reduce the risk of flooding. It was built towards the end of the eighteenth century and is now a haven for wildlife, including the glass eel, whose numbers have fallen to less than 5 per cent of their 1980s population.

The origin of glass eels was a mystery for hundreds of years, as British anglers never caught anything that looked like a young glass eel, and naturalists couldn't work out where they came from. Aristotle even speculated that they were related to earthworms, which he believed grew out of the mud. Scientists have since discovered that freshwater eels carry out phenomenal journeys to spawn in the ocean, travelling as far as the Sargasso Sea to reproduce before returning to English rivers. Like salmon, they swim upstream and therefore have to overcome numerous obstacles to return to our rivers, one of which is the modern sluice gate like those found in the King's Sedgemoor Drain. As part of a scheme funded by the European Union, eel passes are now being installed to help them climb over sluice gates, and in 2008 one was installed on the King's Sedgemoor Drain, on a sluice called the Greylake.

We arrived at the gate, but to my disappointment the sluice was barricaded off by a barbed-wire fence. I knew what the passes looked like and where they would be. Eels are strong swimmers and climbers, and in order for them to move over the sluice, scientists had designed passes that were essentially strips of artificial grass that the eels could slither over and into the next section of water. The passes were hidden at the edge of the sluice, and from our angle we were unable to see them, so we checked we were alone and that the fence was strong enough to hold our body weights, and then climbed around the perimeter of the fence so that we were almost dangling into the water. From there we could look down onto the sluice and see the eel passes. The eels travelled at night, so we didn't expect to see any, but it felt good to see these little strips of artificial grass tucked away, a reminder that wildlife can thrive in the modern world if we just make room for it. Pleased with our mischief, we clambered back before anyone could see us, then headed back to the camper van and started driving to our next destination: Nether Stowey.

We travelled west for half an hour before arriving at the pretty village of Nether Stowey just outside the Quantock Hills, a hill range and Area of Outstanding Natural Beauty stretching out towards the Bristol Channel. We had come to visit another of the National Trust properties I had wanted to see for many years – Coleridge Cottage, where the poet Samuel Taylor Coleridge lived for three years and wrote some of his finest works, including *The Rime of the Ancient Mariner* and Kubla Khan. It is also the gateway to the Coleridge Way, an 82km walk across the Quantocks and

Exmoor, and a favourite rambling route for Coleridge, William and Dorothy Wordsworth who walked extensively through the district and composed much of their *Lyrical Ballads* poetry collection while reflecting on the surrounding landscape. After an afternoon wandering around the cottage and gorging on ginger loaf in the tearoom, we left Nether Stowey behind and headed out for an early-evening ramble along the Coleridge Way. We couldn't quite manage the 82km route before work on Tuesday morning, so instead we set out on a circular section that would take us across the landscape, starting at dusk and finishing under the stars, bringing us back to the warm camper van and the kettle.

We crept out of the village and made our way westwards through tangled conifers whose branches blocked out the darkening sky above us, and after a while the track plateaued on a vast hilltop, carpeted with heather and gorse. No flowers bloomed here; the gorse had been burned as part of a heathland conservation scheme and now stood dark and twisted against a slate sky. It was a beautiful and stark landscape, and one that hadn't changed much since the days when Wordsworth and his friends walked here together. In his poem 'The Thorn', Wordsworth tells the story of 'a woman in a scarlet cloak' who sits beside an 'aged Thorn':

> No leaves it has, no prickly points;
> It is a mass of knotted joints,
> A wretched thing forlorn.
> It stands erect, and like a stone
> With lichens is it overgrown.

Beside the thorn, he writes, lies a 'heap of earth o'ergrown with moss … like an infant's grave in size', and goes on to describe how nobody knows what lies beneath the heap of earth. The woman was once pregnant out of wedlock, and jilted by her lover who ran off with another woman; many believed she 'hanged her baby on the tree', while others said 'she drowned it in the pond' nearby. Nobody knows what happened to her child, but now she sits on the moor all day and night, crying 'Oh misery! oh misery! / Oh woe is me! oh misery!'

These stretches of thorny heathland are one of the main characteristics of the Quantocks, together with thickly wooded slopes and deep, narrow valleys called combes that each contain their own miniature landscapes and microclimates. To the north we watched gulls drift over the Bristol Channel. Nearby, in a marshy copse beside the path, historians believe Coleridge and the Wordsworths took one particular night walk that would later inspire Coleridge's poem 'The Nightingale':

> All is still,
> A balmy night! and though the stars be dim,
> Yet let us think upon the vernal showers
> That gladden the green earth, and we shall find
> A pleasure in the dimness of the stars.
> And hark! the Nightingale begins its song,
> …
> A melancholy bird? Oh! idle thought!
> In nature there is nothing melancholy.
> …
> 'Tis the merry Nightingale
> That crowds, and hurries, and precipitates
> With fast thick warble his delicious notes,
> As he were fearful that an April night
> Would be too short for him to utter forth
> His love-chant, and disburthen his full soul
> Of all its music!

While nightingales were more abundant when this poem was written, their song can still be heard across the south of England, like beads of rain falling onto a lake, a rare and precious voice in our landscape. The nightingale's song, whose name originates from the Old English for 'Night Songstress', features heavily in literary history as one of love and passion – such a beautiful song, in fact, that in the nineteenth century, bird-catchers tried to capture large numbers of nightingales for the caged songbird trade. Most of the birds died quickly in captivity, but some survived until autumn, when they killed themselves by dashing their bodies against the cage bars in an attempt to follow their migration instinct.

Many of the poems referring to the nightingale's song often mistake the singing bird as a female, but it is actually the male that sings. They do this predominantly at night to serenade migrating females who might be flying over, tempting them to come down from the sky and mate with them. A recent study at Freie Universität Berlin found that, although we might find nightingale song beautiful, for the females it is more about decoding the song to discover complex information about family values, and how much support the male is likely to offer its family. The more complicated nightingale songs are much harder to sing, and therefore indicate good physical condition. All of this information can indicate his age, where he was raised, the strength of his immune system and how motivated he is for the family.

We passed the marshy copse and continued to climb, our route ascending to the summit of a barren peak where we discovered a heap of stones piled together by travellers as a passing gesture. We added a couple more to the pile and continued, dropping down into deciduous woodland and stopping suddenly on the path. We could hear something in the half-light ahead of us, like running water but with a beating thump coursing through it: there before us, a herd of red deer started crossing between the trees, alarmed and cautious at our footsteps, seeking safer ground. This breed of deer move around most at dawn and dusk, when they are less likely to be disturbed by ramblers and hikers, and in the day they spend hours 'lying up', resting quietly between feeding bouts.

In the early twentieth century, the Quantocks was a popular destination for stag hunters, but the red deer that lived here then – and still today – were imported from nearby Exmoor specifically to provide game for hunters. As red deer are adapted to colder climates like those in the Highlands, they survived easily in the milder habitats of south-west England, growing much bigger than the other deer that lived here and making them even more of a target for hunters. During rutting season, stags will visit the more boggy areas of the Quantocks' lower combes and roll around in the wallows, coating their fur in peat that will then dry to give them an ominous, black appearance. Naturalists believe they do this for a number of reasons: firstly, to make them look more frightening to the other stags, which will give them an advantage when trying to prove themselves as a dominant male; secondly, by urinating in the mud before they roll, they coat themselves in musky pheromones to make them more attractive to the females. In spring, the females then join the stags in their wallows, as it also helps the males and females moult their winter coats.

We stood mesmerised as the herd slipped away into the woods, and continued our journey through the trees until we reached the road that would take us back to the camper van, past thickets of wild garlic and a sleepy herd of cows who licked my hand as we passed. Soon, the sky became swathed in wisps of cloud and it started to rain heavy, cooling droplets that remained slow and fat, those delicious swollen raindrops that form puddles on your skin and yet never seem to create a proper downpour. We passed a livery, empty of horses who would probably be roaming across the fields nearby, and on the road outside I pocketed a rusted horseshoe lying in the rain.

While landscapes forge and reshape themselves over the centuries, there was something comforting in knowing the paths we walked that evening were once crossed by Coleridge and the Wordsworths 200 years ago. These writers' journals and letters show they liked walking at night, crossing the Quantocks when the paths would have been more worn – not just a place for Sunday walks, but a throughway for traders and travellers. Coleridge and the Wordsworths told the locals they were roaming the hills and 'making studies', but even then their innocent pursuits were misinterpreted. In the midst of the French Revolution happening just across the sea, the local people suspected that these literary visitors had other ideas. They weren't originally from the area, spoke French and held radical views, and residents believed they might be helping with preparations for a French invasion along the Bristol Channel. At one point a government agent was even sent to investigate these rumours, convinced the three writers were searching for navigable rivers through which the French ships could sail inland.

Growing up in the nearby town of Ottery St Mary, Coleridge had already experienced one night in the wilderness, although it wasn't quite the same as the inspirational, educational walks he would later experience with his friends Dorothy and William Wordsworth. According to a letter he wrote to his friend Tom Poole, he recalled that when he was a young child, he had an argument with his brother over a plate of cheese. They ended up in a fight, and when their mother came to break it up, Coleridge fled from the house and ran 'to a hill at the bottom of which the Otter flows', 1.5km from his house. Here he allowed his anger to die away and read a little book of prayers he had in his pocket, before falling asleep and rolling slowly down the hill where he awoke several times 'wet & stiff, and cold'. A search party was dispatched by the family, and he was eventually found by the Ottery town crier, who had heard his sobs and found him lying under a gorse bush. Coleridge recalls how he was carried back home, where:

> I remember, & never shall forget, my father's face as he looked upon me while I lay in the servant's arms – so calm, and the tears stealing down his face: for I was the child of his old age.

In the eighteenth century, people of the poorer classes living in the country would have been accustomed to lower levels of light, living and working outside when they could. Cottages were fitted with small windows, and candles were luxury items made of precious beeswax or rendered animal fat called tallow. Moving about in the darkness would have been something they were more accustomed to, and unlike today, most rural people would have known how to use the night sky to navigate their way across the moor. The sky would have been even darker and more dense than it is today – even in the cities, where light-pollution levels would have been much lower. On a clear night the stars would be bursting from the sky, and if travellers kept a specific star or constellation within sight, it would have helped them keep to the right path or at least move in the right direction.

If the sky was cloudy, the stars might not have helped, but one theory suggests that travellers carried white stones with them that shone faintly in the dark, dropping them along the route like breadcrumbs to help them retrace their steps and find the return journey home. In 2002, environmental artist Andy Goldsworthy created the Chalk Stones Trail, a collection of giant balls made of chalk, a material originating from the skeletal remains of marine creatures deposited 70 to 100 million years ago, when a warm sea covered most of southern England. The South Downs are based on chalky soil, giving them their rich abundance in wildlife, and the Chalk Stones Trail takes ramblers along an 8km route into the heart of West Sussex. At night, the stones are said to glow in the moonlight, marking out the trail just as the dropped white stones would have done for those walking in the past.

Aside from the sense of wonder we gain from looking up at the universe at night, why do we need our skies to be dark? Is light pollution genuinely a problem or is it just something that doesn't fit with our visions of a perfect landscape? It is only in the last 100 years or so that some of Britain's skies have stopped being naturally dark, not just around urban areas where light pollution spreads far into the sky, but also above small towns where residential areas direct large amounts of light upwards. It is these sources of light pollution, and excessive use of artificial lights, that the International Dark-Sky Association (IDA) combats, first founded in 1988 by two astronomers. The IDA is a non-profit organisation based in the United States, with more than 60 local groups around the world that advocate for the protection of the night sky through the reduction of inappropriate and unnecessary light-pollution levels, including light shining directly into our eyes, called glare, and light shining above the horizon, called skyglow.

With climate change, pollution and illegal trophy-hunting making most of the headlines in environmental circles today, protecting the night sky might seem like a low priority on the list of problems to solve. But the truth is that by resolving many of the issues that compromise the darkness of our skies at night, it will lead to benefits for the greater environmental cause. Like most aspects of the ecosystem, the 'health' of the night sky is connected with the health of everything else on the planet.

According to studies by the National Optical Astronomy Observatory in Arizona, poorly aimed and unshielded outdoor lights waste more than 17 billion kilowatt-hours of energy every year in the United States alone. The US Department of Energy estimates that 13 per cent of home electricity usage goes towards outdoor lighting, and with more than one-third of the light produced being lost to skyglow, this means that residents are spending $3 billion on wasted light. In the meantime, around 5 million tonnes of carbon dioxide are being released unnecessarily each year by wasted outdoor lighting, which would mean around 600 million trees would need to be planted to offset these carbon emissions.

Aside from energy wastage, which contributes directly to climate change and global warming, light pollution has been proven to affect the habits and behaviours of various species of wildlife. Many animals, like snakes, salamanders and frogs, restrict their movements when the moon is full in order to avoid predators spotting them in the moonlight. This means they tend to hunt more on moonless nights, but as artificial light pollution is spreading further across the habitats of these species, they are spending less time hunting and more time waiting for the light to dim – and as it's not a natural light with natural rhythms, it never does.

One experiment in Pembroke, Virginia, saw ecologists string a line of white holiday lights along transect lines to test the effects of artificial lights on amphibians that usually emerge about an hour after dusk to hunt for food. They discovered that when the string of lights were turned on, the animals stayed hidden for a whole extra hour, meaning not only were they spending less time looking for food, but by the time they emerged they had lost an hour's worth of food to other predators.

Another research paper found that many species are susceptible to night blindness. In trying to find ways to reduce deer–vehicle accidents in the United States, local authorities had increased artificial lighting along highways, but studies suggest this actually makes it more difficult for nocturnal mammals to avoid collisions with vehicles. Many nocturnal animals use rod cells in their eyes to see in the dark but, when suddenly faced with a rapid change in brightness, they are unable to switch back to normal sight quickly enough and are unable to work out which way to run to escape the vehicle. Their retinas become saturated by the lights and the animals become 'night blind', which is what is often called the 'rabbit-in-the-headlights' effect.

It's not only a sudden change in lighting that causes animals to become disorientated, but the presence of brightly lit areas in their paths. The wild puma, found across the Americas, travels between territories at night in a corridor, but researchers have found that when their path is interrupted by artificial lighting, they will often stop in their tracks and end their journey there. This is because the puma is another animal that is susceptible to night blindness, and when they come across a brightly lit town or industrial centre, they are unable to see the dark wilderness beyond; rather than risk the unknown, they will wait until daylight to move beyond that point and continue their journey.

Birds also use the moon and stars for navigation during their biannual migrations, often travelling at night to avoid detection by predators. According to one story, in 1954 over 50,000 birds died at Warner Robins Air Force Base in Georgia, United States, over the course of two nights, after following the artificial lights and flying straight into the ground. The Leach's storm petrel is another species that is particularly drawn towards lights, spending most of their time offshore feeding on bioluminescent plankton in the sea. These birds can be attracted to lighthouses, offshore drilling platforms and the high-intensity lamps used by fishermen to lure squid to the surface of the water, all of which can result in the deaths and injuries of hundreds of birds.

Along the highly developed coastline of Florida, the beaches are home to a number of rare turtle species, including loggerheads, leatherbacks and green turtles. Females come ashore at night to lay their eggs, but researchers have found their visits are declining, discouraged by the bright lights of human development. Local authorities have asked residents to turn off beachfront lights during turtle nesting season, but this doesn't address the larger problem of skyglow above the cities. When the eggs are laid, the newly hatched turtles are struggling to find their way back to the sea. Usually guided by the stars and moon, hatchlings are starting to crawl back inland towards the artificial lights of the towns, or even crawling aimlessly down the beach until dawn, when they are more likely to be eaten by predators.

In cities, wild birds are shifting the start of their early-morning dawn chorus to avoid light pollution from urban developments, although researchers suspect this could also be due to noise pollution. One study conducted at five airports in the UK found that birds had started to anticipate the morning rush of planes on the runway, changing their song times in order to avoid the noise and make themselves more audible to other birds. As the dawn chorus usually takes place just before it is light enough for the birds to navigate and forage for food, by singing earlier this means they are increasing their efforts without the opportunity to replenish their energy stores immediately afterwards. It also makes them more susceptible to nocturnal predators whose active hours are more likely to overlap with those of the birds.

It isn't only wildlife that suffers when light-pollution levels go unchecked. Recent studies have linked artificial light at night to an increased risk of diabetes, obesity, depression and cancer, as well as numerous sleep disorders. When we don't spend enough time in darkness, our bodies don't make enough melatonin, the hormone that maintains our sleep–wake cycle and helps to regulate the rest of our hormones. According to one study at Stanford University, California, people living in urban areas are exposed to artificial lighting that is three to six times more intense than people living in small towns and rural areas. The study showed that those living in intense light areas were 6 per cent more likely to sleep less than six hours per night, and were also more likely to wake up confused during the night, and feel unnaturally tired in the daytime.

Aside from all the health effects, all the negative disruption to our fragile wildlife, and all the money and energy wasted on excessive artificial lights, perhaps the most effective way to understand the importance of dark skies is to simply look up. Every year, more and more people are losing access to the night sky, unable to see the stars and planets due to light pollution. One of the most well-known galaxies, the Milky Way – that band of sparkling, ethereal light smeared over our heads – is becoming so difficult to observe that a growing majority of people believe they have never seen it. In 1994, an earthquake in Los Angeles knocked out all the power lines, and the emergency services received numerous calls from residents reporting that a strange, silvery cloud had appeared in the night sky. They were seeing the Milky Way as, for the first time in their lives, it was not obliterated by the skyglow emitted from the city.

For the light of a star to reach the human retina, a long, difficult journey must first take place. A star is born when a group of atoms are squeezed under enough pressure for their nuclei to undergo fusion, a nuclear reaction in which atomic nuclei fuse together to form one heavier nucleus, releasing energy at the same time. All stars are created as a result of a balance of different forces. One of these is the force of gravity, which compresses atoms in interstellar gas until the fusion reaction begins; once this has happened, the reactions exert another outward force. As long as the inward force of gravity is equal to the outward force generated by the fusion reaction, the star remains stable, and as soon as the star is formed, gamma ray photons are released from the nuclear core. From here, they travel through the different layers of the star, until they finally burst free and speed through space at almost 300,000km per second – known as the speed of light. Depending on where they originated from, these photons can travel for as little as eight minutes or as long as millions of years before making it to the earth's atmosphere, where they finally reach the back of our eyes and appear as a tiny sparkle in the sky that we then call a star.

It's amazing to think of starlight flying through the cosmos and down to our planet, only to be lost in the skyglow being emitted from artificial lights above our cities. Governments around the world have been trying to reduce light pollution in a number of creative ways, including replacing dazzling, outdated outdoor lights with more advanced, low-glare versions, encouraging the use of motion sensors to switch off lights when they are not needed, and replacing old street lights with downward-facing lamps to reduce the amount of light being projected into the sky.

It will come as no surprise that some of the strongest advocates of dark skies are stargazing and amateur astronomy groups. Back at Butser Ancient Farm, we have a Hampshire-based group who spend the long winter nights huddled between the roundhouses with their telescopes pointing towards the stars, making use of the fact that the farm lies within the South Downs, an International Dark Sky Reserve, and is one of the darkest points in the National Park. There is something mesmerising about amateur astronomy – not just being able to watch the stars and planets, but watching the stargazers themselves. Their passion for observing the night sky is infectious, as is their skill with night-sky photography, their knowledge of telescopic technology, and their awareness of which celestial bodies are lingering above our heads at any given moment. Today, astronomy is one of the few sciences where amateurs can make important contributions through recording their observations all over the world.

More than anything, protecting our skies from light pollution is vital for reminding us of our place in the universe. Somewhere along the way, we accidentally became the dominant species on earth, and with this came a sense of superiority over every other living thing we share the planet with. Many of us now believe the earth is there to serve our human purposes alone, but when we gaze up at the night sky and count every star, every planet lost in space above our heads, we are reminded that we are only one species in the wide universe, and there is still so much we don't know and have no control over. The night sky reminds us of how vibrant our planet is, urging us not to take it for granted, but to protect it and care for it – this living, breathing world suspended in the cold, glittering vastness of space.

TIF

CHAPTER SEVEN

The Mountain

We had been driving for six hours now, the camper van stuffed with two duvets, a crate of food, Bananagrams, binoculars and two now-empty packets of Veggie Percy Pigs from the petrol station. It was the usual set-up for a weekend away in the camper van, but we had long since exhausted the driving playlist and already reached full-on wriggle-mode from sitting down for too long. Phone reception had been lost 20 minutes ago and our coffees had been drained; we could only wait for our destination to appear.

And then the rain started.

There were only light specks at first, the type that makes everyone remark: 'Oo, look. Rain.' The windscreen wipers smeared them into thin streaks mingled with dirt and road dust, but then more appeared, and more, until we were caught in a cyclonic downpour in the dark, driving through the valleys of Snowdonia National Park in north Wales.

Like all our trips in the camper van, we didn't have an exact destination in mind. We were here to explore Snowdonia, another of Britain's International Dark Sky Reserves, spread over 2,130km^2 – around 10 per cent of the total land area of Wales. With such a wild, rugged landscape, the park supported little human settlement, and consequently it was a starlit haven, naturally dark and separated from the glow of the coastal cities. Astro-tourism had been drawing visitors to the park before it was made a Dark Sky Reserve in 2015, but now people came from across the country to gaze up at the stars above the wild Welsh landscape – including us.

We'd come to witness the beauty of the National Park at night, but while we were here we had decided to spend the day climbing Snowdon, something neither of us had yet done. We planned to set off the next day, so with all the organisation we could manage, we typed 'Mount Snowdon' into Google Maps and presumed it was now taking us as close as we could get to the bottom of the mountain without leaving tarmac. Most of our camper van journeys involved heading vaguely in the right direction and pulling over when we were nearly there and came across a pleasant sleeping spot. But it was now late evening on a thunderous night in March, and we could see nothing beyond the edge of the road, where our headlights feebly petered out.

Nevertheless, it was deliciously cosy, bumbling along in the dark with a box of food in the back, surrounded by duvets and with a cup of coffee available at a click of the stove ignition. There's nothing like a camper van to make you feel completely in harmony with the outdoors. When the sun pours down in summer, we throw open the doors and feel the warmth on our skin, reading books beside rivers and fields, painting, and eating crisps. When it's miserable, an irrepressible sense of snugness creeps in, watching the rain hammer against the windows of our little nest on wheels. We call it ours, but technically it belongs to Dave's parents who offer it out to the family all year round in exchange for one of our cars. There's a family rota in the kitchen on which we had scribbled 'Tiff and Dave to Wales', although at this time of year we needn't have worried it would already have been reserved. North Wales in March is beautiful, but there was a reason we had brought along our ski jackets and plenty of blankets.

At this time of year, visitors to Snowdonia can expect rainfall almost every other day and an average temperature of just 6.8°C. It's hardly the Costa del Sol, but when a place is as wild and green as Snowdonia, there have to be some downsides. The land is awash with rust-coloured grassland, moss-strewn rocks and turquoise freshwater pools as clear as glass, and at the centre of it all stands Snowdon, the highest point in the British Isles at 1,085 metres above sea level. The name Snowdon comes from the Old English 'snow hill', presumably for the snowfall spread across its peak, while the Welsh name *Yr Wyddfa* means 'the tumulus', thought to refer to the old folk story of Rhitta Gawr, the strongest and most violent giant of the ancient times.

The story goes that Rhitta became fed up of the humans who fought to rule over the land of Prydain, and as the strongest and bravest of them all, he decided he should be in charge. One after another, he attacked and defeated the leaders of all the kingdoms, shaving off their beards and weaving them into his cloak as proof of his victory. Kings soon came from foreign lands to try and slay the giant, but each one was defeated until, finally, the great warrior Arthur set out with his men to Rhitta's fortress in the mountains of Gwynedd. They fought on the highest peak of the highest mountain, hammered by a winter wind, and as the battle continued, swords were shattered, shields torn and bones broken. By the end, both were wounded and covered in blood, but with one last burst of determination, Arthur brought his sword down deep into Rhitta's skull, and the giant was dead. Arthur and his men piled rocks onto the fallen giant, and the place was named *Gwyddfa Rhita* – Rhitta's Tomb – which changed over time to become *Yr Wyddfa*.

We had now come so close to the mountain that we could see the visitor centre further along up the road on our Google Maps screen, so we decided to pull over at the next suitable place and settle down for the night. The only problem was that the torrent of rain made it impossible to see more than a few metres ahead of us, and without any phone signal I couldn't reprogramme Google to tell us exactly where we were. The map was just coloured shapes – no road names, no directions – but in the end we found a quiet patch of road with a large blue splodge next to it on the map, which we cleverly decoded as water. We pulled up on the verge and put the kettle on, set the bed up and made supper, utterly lost in the dark. At least in the morning we would wake beside some pretty body of water, and that was good enough for us.

The rain threw itself against the windows and walls of the camper van, and inside we were jostled about while we sipped tea and ate beans on toast, cosied up in the newly made bed. Tucked away under two duvets, we felt warm and snug, but deep down we knew we would have to venture back outside before the morning came – we'd drunk too much tea. For now, we settled in with the blinds open, watching the rain fall against the glass and the black nothingness of Snowdonia fade away behind it. We knew it was cold outside – we could already feel it seeping through the walls – and by early morning the temperature would have dropped even lower. But we were visiting northern Wales in early spring, and expected nothing less than a few drops of rain and a nip in the air. We were looking forward to climbing Snowdon in the morning, if only to warm our bones against the cold March weather. At last we finished our tea, ventured out into the fearsome night and returned as soon as we were able, rocked to sleep by the fumbling hands of Wales' thunderous gales.

Despite having ancestral roots in Wales – something my mum is extremely proud of – I was unfamiliar with this part of the country and wanted to explore more of it, particularly as it was one of only a handful of sites across the UK that had been given International Dark Sky Reserve status. Being from the South Downs, where the stars were also bright and undiluted by light pollution, I was keen to see the stars, but it wasn't essential. Instead, I was here to explore a place that was so wild and rugged that it was almost cut off from the rest of Britain, to learn about the communities that lived here and how they had formed their own connection with the darkness.

By morning the storm had passed, and we woke to the peace of the Welsh hills. In daylight our mystery landscape revealed itself – a beautiful reddish-purple moorland interspersed with rivers and lakes, the morning sunlight dancing off the surface of the blue splodge we had parked next to. We brewed a pot of coffee and grilled a couple of pains au chocolat until they were oozing their filling everywhere, then tidied away and drove on to Pen-y-Pass, the car park frequented by hikers who want to ascend Snowdon up the Miners' Track, one of the easier routes up the mountain. It was Friday, which meant there were fewer hikers to share the path with as many visitors would be at work. Dave was self-employed and could take time off when he liked, and at the beginning of the month I had left my job at Butser Ancient Farm to finally take the plunge and go freelance.

I loved my job at Butser and, as everybody always joked, nobody ever really left the farm. I would still volunteer there, cycle over and visit the goats, use the tech pods to develop my wood-carving and metalworking skills, and generally just bask in the beauty of the place. But on the day I left, I had just one week to go before the release of my first book *Food You Can Forage*, and I had already been thinking of taking a week or two off to focus on the promotion. Dave had been encouraging me for a long time to go freelance, but it was only in the time leading up to my first book being published that I finally felt ready to do it. With the momentum of the book I had made lots of contacts, spoken at events and secured commissioned work for magazines, and I was feeling optimistic about the future. It was now or never. I helped welcome my replacement, Rachel (who was far better at the job than me), and everybody wished me well with cards, flowers, a bottle of gin and a beautiful card painted by my ever-inspirational friend Victoria who worked in the upstairs office. On the last day of February I waved goodbye to the farm and started my career as a freelance writer, artist and environmentalist.

There were evident downsides to leaving employment and a juicy salary behind, but I'd never felt so liberated, so free to pursue the life I wanted and to spend my days doing exactly what my creativity encouraged me to do. The best part was choosing how and when I worked, being able to wake up in Snowdonia National Park on a Friday morning, to climb a mountain and write notes about the Welsh wilderness to fill my next book. My income may have been questionable and my future mortgage applications looked dubious, but this was freedom like I'd never known before, and I hoped I could hold on to it forever.

We arrived at Pen-y-Pass and layered up with all the clothes we had, including ski jackets, woolly hats, gloves and hiking boots. The sun was shining and the temperature pleasant at the bottom, but the English hadn't called it Snowdon for nothing. A sign beside the path told us the temperature at the summit would feel like -10°C, with bitter winds blowing in 65kmph gusts. In fact, beside that sign there was another message from the mountain warden, warning that only climbers with crampons and pickaxes would be able to climb to the top of Snowdon today, as the conditions were too dangerous for normal hikers. We were disappointed, but as we were here and ready to ramble, we decided to climb as high as we felt we could, and then turn around and enjoy a leisurely walk back. We packed tomato and avocado sandwiches, trail mix, walnuts, dried mango, chocolate, bananas and a thermos of coffee, before locking up the van and starting the long climb up the mountain.

We had been without signal or WiFi for almost half a day by this point, and it was bliss. In the last few weeks I had made the decision to delete both my Facebook and Twitter accounts – a scary move for somebody who had been on various social-media platforms since the age of 14 and the golden days of Bebo and Myspace. I had also been using it consistently to promote my freelance writing and artwork, and it had been thanks to Twitter that I secured my first book deal in 2016. Social media was a powerful, inspiring tool and it had been hard to release myself from it – particularly

Twitter, which offered such a window onto the outside world. I tried simply restricting my usage but I didn't have the discipline, which is no surprise when these companies invest millions of pounds into making them as addictive as possible. I didn't see social media as an evil entity – merely a product of the society we lived in – and while I could see how useful it was for others, I found it was having less and less of a positive impact on my life. Twitter had become an echo chamber of negativity, full of bickering and clickbait, and one morning I decided that, although I might lose a few book sales here and there, I would be a happier person without it.

I also found I had become much less sociable the more I was consumed by social media. This wasn't simply a case of checking my phone when I was out with friends, though I was guilty of that too. But by being forever updated with what other people were doing, it meant I didn't need to bother asking them about it in real life, and I realised there were hundreds of real conversations I was missing out on by always having access to a summarised version of my friends' lives. As I started to wean myself away from my online persona, I was genuinely excited to see people at the pub, to text and call friends I hadn't seen in a while and to find out what they were doing or how they were feeling. And as I spent less and less time on my phone, I found time for the other things I loved but that were always being superseded by my phone: I finished knitting the blanket I had started months before, I read more books, watered the plants, watched more documentaries and followed the birds around the garden with my binoculars – all the lovely things we do to break up the more important parts of our day. Slowly, minute by minute, I was freeing myself from social media, something we are told is a necessary part of life, an essential tool, a form of entertainment we cannot live without. But the truth is it's just that – entertainment. Not love or friendship or art, but just a reflection of those things. And while a reflection can be a wonderful window into our lives, it isn't real life itself.

By the time we drove to Snowdonia that weekend, the only social-media platform I had left was Instagram, which I found to be a more positive and inspiring online place to spend time. I decided to give Instagram a chance and see if, by having only one platform to focus on, I could be more present and enjoy the real world more. If it didn't work – if I became just as enslaved to Instagram, just as obsessed with instant, positive feedback and statistics and keeping up a persona that wasn't real – then I was prepared to give that up too, and give myself back the hours of time I spent on my phone every day. At least here in Wales I didn't have access to anything online; without a signal I could only enjoy the scenery, the smells and sounds of the mountain, and the hours of real conversation with the man I loved.

I'd brought my snazzy bridge camera to take photos of the birds and scenery along the way, so my phone stayed quietly stowed in my pocket with only its GPS tracker working away to record our route and mileage. The phone itself had been on silent for about eight years as I hate ringtones, so we soon forgot all about it, focusing instead on the beauty of the place, turquoise lakes sparkling in the spring sun, quartz rocks and luscious greenery, birds hopping over the paths and into rainwater puddles that had formed in the rocks overnight. We saw a wheatear washing himself: a handsome, ground-dwelling bird with a bluish-grey back and a sharp, black cheek stripe outlined in white. This was the first I'd seen of the year, as they migrate over from Africa in the spring, another species to fly through the night, relying on their fat reserves for energy.

The Miners' Track climbed across the base of Snowdon and wrapped itself around the mountain like a snake, the steps crumbling, hikers stumbling over craggy stones and boulders. The path felt old – not just in the sense that mountains are geologically old, but with a taste of human history too. It was built to serve the Britannia Copper Mine on Snowdon, but before that the miners had had to heave the copper up the eastern side of the mountain so it could be taken down to Llyn Cwellyn on the other side by a horse-drawn sledge, where it was then transported on to the town of Caernarfon to be processed and dispatched.

The first recorded mine on Snowdon dates back to the 1800s, but there are rumours of copper mining here in the Roman period. In the mid-eighteenth century, demand for copper increased when warships started being built with copper bottoms to prevent worms boring into the wood, and a mine was opened at Cwm Dyli, where the hydroelectric power station now stands. Copper extraction on Snowdon was never a lucrative endeavour, with many mining companies going bankrupt and most closing for good by the start of the First World War. The mountain is still speckled with fragments of mining history: an abandoned barracks for the miners, the ruined crushing mill, crumbling buildings and even subterranean sections of the mine that can be explored with an experienced guide. The Miners' Track took us past a lake called Llyn Llydaw which, although beautiful, we had heard was ecologically almost barren, a result of contamination from the shafts of the old copper mines. On the bank overlooking the lake stood the remnants of the abandoned mine, a haunting, broken building jutting out of the mountain like it had been hewn from Snowdon itself, staring morbidly across the blue lake.

Travel back a few centuries, and this is where legend claims King Arthur came across the Lady of the Lake one night. Given powers of enchantment by the goddess Diana, she enticed Arthur and gave him his sword, Excalibur. Alfred Lord Tennyson reimagined the scene in his poem collection *Idylls of the King*, in which Arthur reflects on his first visit to the lake:

'Take my brand Excalibur,
Which was my pride: for thou rememberest how
In those old days, one summer noon, an arm
Rose up from out the bosom of the lake,
Clothed in white samite, mystic, wonderful,
Holding the sword – and how I row'd across
And took it, and have worn it, like a king.'

As he lies wounded, Arthur asks his marshal Sir Bedivere to return Excalibur to the lake in order to fulfil the prophecy etched onto the blade. On his third attempt, hypnotised by the beauty of the bejewelled sword, Sir Bedivere manages to throw the sword into the water and recalls how he saw the Lady of the Lake for himself:

'Sir King, I closed mine eyelids, lest the gems
Should blind my purpose, for I never saw,
Nor shall see, here or elsewhere, till I die,
Not tho' I live three lives of mortal men,
So great a miracle as yonder hilt.
Then with both hands I flung him, wheeling him;
But when I look'd again, behold an arm,
Clothed in white samite, mystic, wonderful,
That caught him by the hilt, and brandish'd him
Three times, and drew him under in the mere.'

It was easy to imagine an enchantress emerging from the lake, but the abandoned mine brought darker thoughts into my head. I remembered learning about mining communities on Victorian Day at primary school, when we'd all dress up in smocks, play hoop-and-stick, and have sepia photos taken. We were told how entire families were encouraged, out of poverty and the need to survive, to work in the mines, although I later found out that coal mining was more associated with child labour than copper and tin. Women were expected to carry the same loads as men while working for less pay (can you imagine?), and to continue working into the last days of pregnancy. Both women and children, who were smaller than their male colleagues, were required to crawl through tunnels less than 60cm high for 12–14 hours a day, hauling wagons for coal, tin and copper, to which they were attached with chains. The air quality was also low in oxygen and full of carbonaceous particles from blasting and mineral dust, with the air in one Cornish mine in Gwennap described as so thick with powder smoke that the workers could barely see their own hands. It wasn't until 1842, after years of campaigning by the Victorian philanthropist Anthony Ashley-Cooper, 7th Earl of Shaftesbury, that the Mines Act was passed, prohibiting the employment of women and children under 10, although the working conditions and long hours were still terrible for the remaining employees, many of whom died in accidents or through severe lung conditions.

We departed the abandoned mine and the lake and continued hiking up the Miners' Track, the air cooling as we walked, the wind bristling on the back of my neck. It was not long before the ground became smothered in patches of snow, thin and watery at first but soon forming thick, velvet slabs, smooth and untouched by footprints or pawprints. We greeted other hikers on their way back, and most told us smugly that we'd never reach the top. Nevertheless, we persisted, and soon came upon two men in their sixties who confirmed that they had been to the summit that day, laughing that if they could do it, so could we. We were delighted! Ever optimistic, I was always more interested in what *could* happen rather than what *should*, and if these healthy sexagenarians could clamber up there, we would too. With renewed vigour, we kept hiking until the path steepened, and was soon lost under the snow. We half-climbed, half-crawled our way to the summit, every step ascending into colder, bitter winds. The path had disappeared behind us, and we could now only traverse over slabs of rock, sheets of snow and glassy layers of ice that made us slip and stumble up the mountain. Eventually, we pulled ourselves up the final shelf of snow and rock to see the mountain railway, closed until summer, abandoned in the early spring weather and glowing copper-red against the snow. We followed it up to the peak and there, 1,085 metres above sea level, a trig point stood brave and firm against the weather, the beacon of hope for every hiker in Britain. It was built on a platform encircled by two flights of stairs, and, battling the wind, we climbed carefully to the top and looked out over the mountain into a panorama of ice and mist.

It was still broad daylight – at least, further down the mountain it was. Up here, the weather was so ferocious that the sky was opaque, the colour of slate, bearing down on us like the vengeful spirit of Rhitta Gawr himself. The rest of Snowdonia might not have been as tempestuous as the mountain summit, but it reminded me of how wild this place was, and how to live here must be to surrender yourself to the power of the elements. Nevertheless, the journey here had been mesmerising, the landscape of Snowdon encapsulated in an enchantment of its own, and I understood why Charles Darwin had written in 1835 that this mountain was 'more beautiful than any peak in the Cordillera' of the Andes. We caught our breath, took a swig of coffee and enjoyed one last look at the bleak, brave beauty of the mountain, before leaving the trig point platform and starting our long descent back to the ground.

This was easier said than done. As most hikers know, the climb down is often more difficult than the climb up, a lesson I learned the hard way when I completed the Yorkshire Three Peaks challenge with my friends Sacha and James: by the time we had finished descending the final peak, our knees were obliterated, and we found ourselves sitting in a traumatised huddle in the Golden Lion pub in Horton-in-Ribblesdale, eating lasagne in silence.

The path down from Snowdon's summit was hidden by snow and ice, so rather than trying to inch down at a snail's pace, I decided to traverse the slope sitting down – a speedy but chilly method that required me to make a snowplough with my legs to stop me falling off the edge and, at one point, crashing into someone's dog. We soon made it down past the snowiest peak and back into the warmer, greener plateaux of the lower mountain, where we decided to take the Pyg Track back for a slightly rockier route home.

By the time we arrived back at the camper van it was late afternoon, and the car park was almost empty. We warmed up over soup and cider, and plotted a short route through the park to one of the waterfalls the area had become famous for. Our layers shed and our bellies full, we started up the camper van and left Pen-y-Pass, winding slowly through the glittering landscape now glowing with a few streaks of sunlight filtering through the afternoon clouds. On the way to the waterfall we passed through a vast, desolate moorland, empty of buildings or people, just miles and miles of burgundy heather that had not yet bloomed with the mauve flowers of a new summer. It was so beautiful, so still and golden in the evening light that we decided to abandon our waterfall plans and sleep here – a night on the moor with nothing but the wind and rocks for company. We parked up and made tea, the sunset already fading and the downy light of dusk settling over the moor.

In a parallel universe, this would have been an ideal place to find a bird of prey I had only ever seen once, flying across the road on the Isle of Mull in the Inner Hebrides of Scotland. The hen harrier is one of the most persecuted birds of prey in Britain, with fewer than 1,000 breeding pairs remaining in the UK – and only four in England. For most of Britain's history they were widespread throughout the country, but during the nineteenth century they were driven to the brink of extinction by persecution from gamekeepers – which was then legal – and dramatic changes in land use. After the Second World War, hen harriers spread out from their dwindling populations on Orkney and the Hebrides and recolonised their upland ranges, aided by new legislation that protected wildlife, a downward trend in gamekeeping, and the replanting of forests. Despite this, hen harrier numbers are still a fraction of what research suggests they should be, and most illegal persecution is now connected with population control associated with the management of grouse moors.

The clue is in the name – hen harriers are natural predators for red grouse, which are bred in large numbers to feed the demand for the grouse-shooting industry, a lucrative pastime that can lead to hundreds of birds being shot each day. I can't bring myself to call it a sport because in sport both sides need to know they are playing and to have an equal chance at winning. The more birds available, the more shoots the landowners can charge for, so to keep numbers up they rid the land of every predator they can, including foxes, corvids, stoats and weasels. And while it is illegal to shoot any bird of prey in Britain, year after year satellite-tagged birds go missing around grouse moors; some are found shot, poisoned or trapped, while others disappear off the face of the earth entirely. The RSPB has even captured video footage of gamekeepers putting out plastic decoy birds to draw others in before trapping and killing them. As a general rule the shooting industry condones raptor persecution, and most agree that it is only the minority of 'bad apple' gamekeepers who are causing the damage. Some suggest it is a mindset that's handed down between generations of gamekeepers, and that shooting birds of prey is just a natural part of grouse-moor management. If this is the case, how can that kind of behaviour survive in a modern, progressive Britain?

In many ways it's a complicated issue, and in others it's not. Grouse-shooting is a rural industry that provides employment for lots of people. There is also proof that grouse moors can be thriving habitats for other wildlife, such as curlews and other ground-nesting birds, although this is hardly surprising when most of their predators have been shot. Some claim grouse-shooting is 'traditional', which is often an easy way to justify barbaric and outdated actions such as fox-hunting, slavery and gender inequality. Take away the economic benefits, the love of tradition and the eerie pleasure that comes from killing things, and the bare bones of the situation are this: hen harriers and other birds of prey, which have been living in harmony with the British landscape for thousands of years, are being illegally persecuted to the point of extinction in order to keep more red grouse alive so that they, in turn, can be shot for 'sport'. If there is logic there, I can't understand it.

To top it off, many grouse moors are also intensively burned – far more than the heathland burned for normal conservation purposes – to engineer the perfect habitat for red grouse. This exposes precious peatland that provides a home for wildlife, helps clean water, prevents flooding, and stores carbon. When it is disturbed and exposed to the air, the carbon stored in the soil decomposes and releases carbon dioxide into the atmosphere, contributing to climate change. Research carried out by the RSPB suggests that 27 per cent of the UK's blanket bog has already lost its peat-forming vegetation due to over-burning, and if we don't restore these upland peatlands soon, carbon-dioxide emissions are expected to increase by 30 per cent with every 1°C rise in average global temperatures.

Standing outside under the evening sky, staring out over the tranquillity of the moor, it was difficult to imagine birds being shot from the sky, but that same week I had read a news report about a satellite-tagged bird known as Aalin, born on the Isle of Man the year before, that had stopped transmitting just east of us in Wrexham, north Wales. I scanned the horizon for signs of life, but I knew I would never see a harrier here, despite their bright, pale feathers and hypnotic flight displays that have given them the nickname 'Skydancer'.

We opened the sliding door of the camper van and cooked dinner with the moor laid out in front of us, our lights being the only source of electricity we could see. It grew darker, until the individual tufts of heather merged into one swathe of red, an ocean of moorland surrounding us on all sides. We had only been there half an hour or more when, suddenly, in the distance I saw a shape. It was a bird, soft and ethereal, floating down like a spirit across the landscape, but it was so far away that I couldn't tell what it was. The binoculars were almost useless at this level of light, and I could do nothing but stare at the bird, willing it into a shape that I could recognise. Finally it drifted closer, searching for something in the heather, and I realised what it was. Certainly not a hen harrier, but another species that is so associated with sharing its habitat that the two birds are known for harassing each other into dropping their prey, a behaviour known as kleptoparasitism.

It was a short-eared owl, gliding low across the heather, his talons almost brushing the tips of the foliage, wings spread wide and yellow eyes focused on the ground beneath him. Even in the half-light, I recognised his bulbous head and dark stripe on each wing flashing as he soared past and away again, off across the moor. After a few metres he landed and disappeared, before erupting again out of the heather a few moments later, back to his hunt, scouring through the undergrowth for something alive. Under only the light of the night sky, he was hauntingly pale, a ghostly vision across the moor like a creature from a Welsh folk tale.

I love folk tales. I remembered hearing about the *cyhyraeth*, a hideous spectre from Welsh mythology whose name literally means 'flesh-wraith'. With withered arms, rotten teeth, tangled hair and corpse-like features, she is thought to roam the Welsh wilderness at night, approaching the window of a person about to die and calling their name before releasing a disembodied moaning sound three times as a warning that their life is coming to an end. She is sometimes conflated with the myth of the *Gwrach y Rhibyn*, another banshee-like figure who shrieks into the night: *Fy mhlentyn! Fy mhlentyn bach!* meaning 'My child! My little child!' Legend claims that she haunts Pennard Castle, a twelfth-century ruin overlooking the sea and built in one night by a sorcerer trying to escape the invading Normans. The *Gwrach y Rhibyn* wanders the castle grounds at night, dragging her rotting body through the ruins, cursing anyone who dares sleep within its walls.

Another myth that haunts the Welsh wilderness is that of the *Tylwyth Teg*, a race of fairy folk that lived in forests and caves near running water. Some of these fairies were benevolent; others were cruel and cunning, thought to have ugly and contorted bodies and faces. One story tells of how the *Tylwyth Teg* harboured an obsession with blonde-haired, blue-eyed babies, and stole them from their homes in exchange for a changeling baby. At first these appeared identical to the stolen child, but as they grew they became shrivelled, bad-tempered and nasty. When one woman's child was stolen she sought the help of a magician, who instructed her to do three things. First, she was to remove the top of a raw egg and start stirring the contents, which made the changeling mutter in an otherworldly tongue. Next, she had to go to a crossroads at midnight during a full moon, in order to spy on the changeling taking part in a secret fairy gathering and confirm he was not her real son. Lastly, she was to buy a black hen and, without plucking it first, roast it over an open fire. When the last feather dropped off, the changeling disappeared and her son was returned, safe and well.

I don't usually believe in ghost stories, but standing there on the edge of the moor, watching the short-eared owl circle in silence over the heather and grass, it was easy to imagine changeling children and spectral, rotting figures staggering down the track towards us. It was an eerie place, and far removed from the towns and cities that surrounded it. After some time, the owl disappeared over the heather, away to new territories where it would continue hunting for prey. Then it was entirely dark – no lilac light of dusk to cast a faded glow over the landscape – just blackness, and from the door of the camper van we could see nothing at all. Even the stars and moon were out of sight, hidden by a blanket of cloud that had settled over every inch of sky several hours ago and refused to move. There was nothing at all to look at but darkness, yet I could still feel the presence of the moorland stretched out in front of me. I could hear the insects murmuring in the long grass, a light wind sweeping over the rocks and up to the mountain on the horizon, and the thought of what might be waiting in the shadows lifted the hairs on my arms. I shivered. The bright lights of modernity hadn't yet penetrated Snowdonia, and in their isolation, away from the rest of the world, the inhabitants had conjured up their own truths and histories, their own interpretations of the human experience and what lay beyond it.

It started to rain. The camper van roof became a drum skin, droplets falling in patterings, the only sound we could hear. We closed the door and I sat at the open window instead, listening to the rain fall across the heather as it released the scent of damp earth into the air. It was peaceful, but then I thought of the *cyhyraeth* moving across the moor, somewhere beyond the window frame, just out of sight. Suddenly I felt exposed, and I imagined a maniac running at me in the dark, out of the wilderness. I thought of the wolves that must have lived here once, and wondered what I would do if I heard one howling on the horizon. How would I feel if I suddenly heard an animal that I knew was extinct in Britain, crying out into the night nearby? It's an unmistakable sound – you couldn't convince yourself it was a fox or an owl. How would the brain reconcile such a noise here in the middle of twenty-first-century Snowdonia?

In a flash of paranoia, I slammed the window shut and drew down all the blinds, blocking out the moor and the secrets it kept. I retreated to the bed, buried myself under the duvets with Dave and a fresh cup of tea, and pledged not to look out at the moor again until dawn and the reassuring glow of sunrise.

CHAPTER EIGHT

The Wickerman

Early May, the dusk was falling and the air around me was pregnant with the aroma of blackthorn flowers that had twisted themselves deep into the hedgerow. I had come back to Butser Ancient Farm, as I did every May, and I was now mildly inebriated, cradling a large pint of cider while I stood beneath a 9-metre wickerman looming into the sky. He stared, mighty and cantankerous across the gloaming, his thick limbs made of hazel hurdles lashed together, with red cedar shingles laid out in rows to form chainmail over his chest. Propped against his arm was a black axe that, the week before, I had painted with a serpentine pattern that was once excavated from some artefact in a Danish tomb.

Behind it all, the dwindling light cast shadows on the forest that bordered the land, home to a nest of buzzards and their mewing young. Between the trees lay a carpet of wild garlic that had begun to unfurl its creamy-white flowers, their aromatic presence floating out of the woods, heralding spring. When we walked in those trees at night, the darkness settled on everything but the flowers and they seemed to glow like stars, a reflection of the sky above the canopy. Once I set my trail camera to film the entrance to the woods, intrigued by what might emerge from the trees. I knew there were badgers and foxes that made their home there, badger setts composed of interlocking tunnels and nesting chambers that might have been passed down over centuries, all hidden away beneath our feet. I caught nothing for days, and then, one blustery night in the early hours, a badger appeared from the gloom – a huge, lumbering beast of a creature with a glint in each eye. He stopped, sniffed at the air, his head turning in every direction until he thought it safe to proceed. Out he trotted, away from the cover of the trees, and stopped again. Sniffed. Ears twitched. Sniffed. Flinched. Startled. Retreat! Back he trotted, into the woods, enveloped by grainy darkness on my computer screen.

There are around 290,000 badgers in Britain, which seems like a lot until you hear that 45,000 are killed on our roads every year. In the 1970s, wildlife protection groups lobbied parliament to make it an offence to attempt to kill, take or injure badgers or to interfere with their setts without a licence. These laws are now contained in the Protection of Badgers Act 1992. Despite this, for the last few decades a number of reactive and proactive badger culls have been launched as part of government initiatives to reduce the spread of bovine tuberculosis (bTB), a highly infectious disease that devastates thousands of beef and dairy farms every year. In 2003, a series of independent trials revealed that reactive culling, in which badgers were culled in the areas where bTB was already present in cattle, actually resulted in a 27 per cent increase in bTB outbreaks compared to those areas where no culling took place. In 2005, a second series of trials then revealed that proactive culling, in which badgers were culled in strategic areas without cattle infection, reduced outbreaks of bTB by 19 per cent within the cull zone but increased them by 29 per cent within a 2km radius. This was because the culling led to changes in the badgers' behaviour, increasing infections within the colonies and leading to the migration of infected badgers away from their usual territories.

In the meantime, wildlife groups had already been searching for alternatives to destroying a protected species and casting the ecosystem into potentially devastating imbalance. In the last few years badger-vaccination programmes have seen success in partnership with vets, farmers and landowners, particularly as scientific research has now shown that cow-to-cow contact – rather than badgers – is the primary cause of the spread of bTB in cattle. Organisations like the Wildlife Trust have been petitioning the government to develop a cattle vaccine and to invest in badger vaccinations, which cost significantly less than the current culling scheme and recognises the fact that 83 per cent of badgers culled between 2002 and 2005 didn't even carry bTB in the first place. Independent experts appointed by DEFRA have been put in place to assess the effectiveness, humaneness and safety of the culls, and deemed them 'ineffective' and 'inhumane', with no significant improvement and further failures year on year.

Despite the science, logic and humanity opposing the badger cull, in 2013 DEFRA announced a further cull in Gloucester shire and Somerset amid intense opposition, with the areas increased in 2015 and 2016 to include seven new licences across Cornwall, Devon and Herefordshire. According to a report in the *Independent* in 2015, the cull is costing taxpayers £6,775 per badger killed, with almost no reduction in bTB outbreaks in Britain, and the often-inhumane killing of 20,000 badgers in 2013 alone.

The European badger is a keystone species, which means it plays such a vital role in the ecosystem that without it, that entire ecosystem could change or collapse with catastrophic consequences. It happened when we removed all the wolves from Britain around the seventeenth century. Wolves were considered vermin, savage killers and a threat to society, and they were persecuted for centuries until the last wolf is believed to have been shot in the Scottish Highlands in 1680. Fast-forward 350 years, and there are now 350,000 red deer roaming through our commercial forests, damaging young trees, crops and habitats. These are animals that would have once been preyed on by our native wolves. Official reports from the Forestry Commission, DEFRA and the University of East Anglia suggest that the cost of damage caused by deer to plantations and commercial woodlands is around £4.5 million a year, while the cost of damage to crops is around £4.3 million. Around 8,000 hectares of woodland with Site of Special Scientific Interest status is currently classed as 'unfavourable' or 'recovering' due to deer impacts, and there has been a 50 per cent decline in woodland bird numbers where deer are present, as the deer eat shrub layers that are crucial habitats for nightingales, willow warblers, chiffchaffs and blackcaps.

The current rewilding debate, in which wolves might be reintroduced into Britain, is a complex one (although I like the idea in theory). But before discussing which animals to bring back into our ecosystem, it is imperative that we recognise the importance of the ones still with us. The European badger feeds primarily on earthworms, but they also help to control populations of mice, rats, squirrels and rabbits, as well as feeding on apples, pears, plums and elderberries and helping to distribute their seeds. Their shaggy fur is also perfect for distributing the hooked seeds of plants like goosegrass and burdock, and they eat competitive plants like the bulbs of wild garlic that might otherwise spread out of control. Badger dung will host its own delicious cocktail of invaluable invertebrates, and as badgers use the same trails over and over again to move between territories, these miniature pathways may create microclimates for butterflies and other pollinators. Their intricate underground setts will also be home to a number of other small creatures, who in turn provide sustenance for their own corners of the ecosystem.

It is impossible to predict how the removal or decline of such a species might affect our environment, but when 60 per cent of our wild species have declined in the last 30 years alone, according to the 2016 State of Nature report, can it be ethical or intelligent to eradicate anything that has a natural right to be there? The modern badger, *Meles meles*, is thought to have evolved in the early Middle Pleistocene, existing within the British ecosystem for thousands and thousands of years alongside other species. In the meantime, agricultural practices have intensified, more land has been taken up for farming, and badger nutrition and habitat is constantly shifting. Is it actually true that badgers need to adapt, be culled, persecuted and disappear from our woodlands? Or is it more likely that, through our own habits, we are making our ecosystems unworkable for ourselves and other creatures?

Tonight, on Butser Ancient Farm as our badgers started to creep out of the cobwebbed earth, I was celebrating Beltain – or Beltane – one of the Gaelic fire festivals that takes place midway between the spring equinox and summer solstice. According to ancient lore, Beltain was a celebration of life and fertility, hailing the start of summer and warmer days to come. It was said that if you bathed in the first dawn dewdrops of Beltain, your year would blossom with beauty and youthful spirit. On the night in question, the Celtic people would traditionally build two large fires using wood from the nine sacred trees: birch, rowan, ash, alder, hawthorn, oak, holly, hazel and willow. Herds of startled cattle were then driven between the two flames as part of a cleansing ritual, purifying the beasts and promoting their fruitfulness. Thousands of years after our ancestors first celebrated these ancient rituals, the fire festival continued in this small pocket of England, a curious event reincarnated for those who wished to forget the turbulent jumble of modern Britain and slip back into the primitive shadows of our past.

The wickerman was thought to be another Beltain tradition, made famous by those grisly horror films starring Edward Woodward or Nicholas Cage (depending on your taste in cinema). The concept of a giant man made of timber and straw was first recorded by Julius Caesar in his *Commentary on the Gallic War* around 58BC. He claimed ancient Druids used the giant effigy to perform human sacrifices to appease the gods, although Caesar wasn't exactly balanced in his view; the Romans saw the Celts as barbarians and regularly exaggerated their behaviour. In truth, there is no archaeological evidence to suggest the Celts carried out these sacrifices. The flames most likely represented the same cleansing force behind the rest of the festival, believed to welcome in the summer and encourage abundance at harvest time. Now, as then, the burn was a sight to inspire and enthuse the crowds as the long, dark winter came to a close.

The evening had drawn in, and I settled down shoeless in the grass, feeling the warm blades between my palms and toes. To my right, Dave watched the titanic figure of the wickerman, and to my left, a city worker and his family sat together on a blanket, exchanging hastily drunk macchiatos on the Piccadilly line for a weekend in the countryside. They'd dined on hog roast and danced with morris men, and now they were waiting for the great display. A few metres away, children fed dusty pellets to a flock of sheep, while a clan of neopagan friends watched the darkening sky with delight, their wrists adorned with charms and pendants. It was time for the burn.

Deep in the base of my stomach, I sensed the drums begin. They started slowly, drawing the gaze of the crowd over to the great man who had just a fragment of time left on earth. Lighting the wickerman was a task allocated by raffle, and this year's winner was a 10-year-old boy. As the drums quickened, he approached the paddock and was passed an Indiana-Jones-style flaming torch. With fear and elation he stepped forwards into the shadow of the giant figure. The torch extended and flames spread greedily, silent as a serpent's tongue, and within seconds the wickerman was alight, and I smiled in the way only humans can in the face of such destruction.

As the blaze engulfed his feet and legs, the drums beat faster in frenzy. Tangerine ribbons of fire rippled into the sky like a whip, and soon the darkness above our heads was glowing with a 9-metre inferno. A veil of ash swarmed through the air and the drums beat louder. Cheers and roars erupted from the masses huddled around the wickerman; beer was drained from tankards, and faces were lit with an apricot bloom as the fire climbed higher. Soon, the body began its inevitable collapse. Those thick hurdles, once so hard and strong, started to crumble in the heat; one leg broke, and the Saxon man fell. Hot sparks showered across the crowd and a cacophony of joy escaped into the night sky.

I glanced around at the landscape, black beyond the flames. The fields and farmlands carved by the Celts over 2,000 years ago still held their shape, and the ancient woodlands from which the wickerman was built had stood for centuries before us. I grasped my cider tightly and stared up at the sky, stars shining through the cloak of hot ash drifting above us like some powdered aurora. Tomorrow, we would all return to our modern lives, our commutes and computers and plastic packaging. But for now I was just another star in the wide universe – and I wanted another cider.

Beltain is one of four Gaelic festivals in the ancient calendar, all of which hang on the natural rhythms of the season, the rising sun, the stars and glowing moon. Later in the summer, the festival of Lughnasadh marks the beginning of the harvest season, traditionally celebrated on the first day of August, with feasting, religious ceremonies, athletic contests, matchmaking, trading and visits to holy wells. The name originates from the Irish god Lugh, a young warrior associated with arts and crafts, truth and the law, the sun, storms and sky. According to *Lebor Gabála Érenn*, a collection of writing detailing the history of Ireland up to the Middle

Ages, Lughnasadh takes some of its mythology from an athletic competition founded by Lugh as a mourning ceremony for the death of his foster mother, Tailtiu. The competition included the long jump, high jump, spear-throwing, boxing, archery, wrestling, swimming and chariot-racing, as well as less physical contests like singing, dancing, storytelling and jewellery-making. Archaeologists also believe that on the evening of the games, a mass arranged marriage would take place in which couples met for the first time, and were given one year and one day to request a no-strings divorce.

Of the four Gaelic feasting festivals, Lughnasadh is perhaps the least well known because, although the authorities of a progressively Christian nation did not object to a celebration of the harvest season, the dates shifted and harvest festivals became more associated with early September. With the autumn came Samhain, another Gaelic festival and one that is perhaps the best known in modern society for its likeness to the popular western celebration of Halloween. A festival to mark the end of the harvest season, Samhain took place on 31 October and 1 November, when the cattle were brought in from pasture and some sent to slaughter. A time of liminality, this was when the boundary between our world and that of the dead was at its thinnest, which meant that the spirits known as *aos sí* could pass between the two. Like Beltain, special bonfires were lit to purify the cattle and evoke protective, cleansing powers from the spiritual world, and the souls of the dead were thought to revisit their old homes for a place at the feasting table. Mumming and guising were common practice, in which people visited their neighbours dressed in costume and recited verses in exchange for food, and games were played involving divination, apples and nuts.

While Samhain dates back to the Neolithic period in Britain, you don't have to look far to find other cultures celebrating a 'Day of the Dead' in one form or another. Perhaps the most famous is in Mexico, where the public holiday *Día de los Muertos* is a time for honouring the deceased by building private altars called *ofrendas*, decorating graves with sugar skulls called *calaveras* and bright Aztec marigolds, and inviting the spirits of the dead to spend the night with their families on earth. In Cambodia, Buddhist families gather together to celebrate *Pchum Ben*, a religious holiday on which people visit temples to remember the dead, offering fresh flowers and foods like sweet sticky rice and beans wrapped in banana leaves. Meanwhile in Hong Kong, the 'Hungry Ghost' or *Yulan* Festival takes place on the 15th day of the seventh lunar month, when many people in East Asia believe the spirits get restless and begin to roam the mortal world. During the festival the spirits are offered both the food and money they might need for the afterlife.

Human civilisation evolved separately in Britain, Mexico, Cambodia and Hong Kong, yet the same patterns appear again and again, connecting our cultures together. Today, we still celebrate ancient festivals repackaged as religious or cultural events. Cynics of the world grumble about the commercialisation of Halloween, but the central idea has been the same for hundreds of years: dress up, eat sweets, get spooky. When I was in primary school my parents let me have an awesome Halloween party at home every year, complete with a range of low-budget costumes, pumpkin soup, 'blood bites' (jam sandwiches) made by my mum and a spine-tingling movie marathon of *Hocus Pocus*, *The Witches* and possibly *Casper* if we hadn't fallen asleep by then. One year, my dad rigged up a line of thread to the fireguard and pulled it over when we were thoroughly engrossed in the telly, causing a roomful of eight-year-old girls to erupt into supersonic shrieks. My parties were the best.

Samhain has almost disappeared for many of us, and with it the act of celebrating the dead. We see other cultures around the world taking part in festivities that we might call morbid, but these are simply different ways of processing death. Today, especially in the name of mental wellbeing, we are encouraged to grieve over the death of a loved one, to take our time to accept somebody has disappeared from our lives, to cry and talk about the loss until we at least get through the initial shock. But essentially, we are encouraged to move on – to visit a cemetery once in a while and keep photos around the house, but to carry on with our lives and focus on the living world. Does this make us better or worse at coping with death? Other cultures celebrating the dead and remembering them with such vitality, seem to focus not on their absence but on the happy memories they created. The art of celebration seems to blur the line between this world and the next, to reduce the concept of death down to just another part of life, an experience to be shared between generations.

At the darkest point of the year comes the last in the cycle of seasonal festivals celebrated by the people of ancient Britain. Imbolc, also known as St Brigid's Day, marks the very beginning of spring and is held at the start of February, around halfway between the winter solstice and spring equinox. It was a time to let go of the past and look to the future, clearing out old possessions, thoughts and habits, making space for new beginnings in the home and in the mind. It's easy to see where the modern tradition of New Year's resolutions comes from, combining the energy of a new seasonal year with ideas of regrowth and regeneration to make us better versions of ourselves. Traditionally associated with the start of lambing season and when the blackthorn tree blossoms, it was a slightly more reserved festival promoting the hearth and home, often involving special fires, feasting, divination and the burning of candles for purification, evoking the power of the returning sun.

Saint Brigid was worshipped in both Celtic and early Christian communities, and was thought to visit virtuous homes on the night of Imbolc. Although the mythology surrounding the festival remains unclear, it is believed that people would lay rushes or hay on the floor of their homes as a bed for Brigid to make her feel welcome, along with food, ale, candles, and a white birch wand she could use to kick-start the growth of new life.

For the people of ancient Britain, Imbolc must have seemed like a light in the darkness. True, early February is hardly the optimum time to celebrate warmth and new growth, and in reality February is possibly the most depressing month to experience in Britain. Short, bleak days and nights are drenched in mist and drizzle, endless and drab, like the world has been painted over with a grey wash. It is relentless. No wonder so many people choose the beginning of the year to escape abroad, abandoning our shores for a burst of winter heat. For our ancestors, however, living without twin-engine jets and all-inclusive hotels, there was no escape. To celebrate Imbolc was to focus on the beauty of the season: the rotund joy of new buds emerging on leafless branches; a change in the air as the decaying aroma of autumn slowly shifts into the freshness of spring; the warmth of pregnant ewes carrying new life in their bellies. In fact, it is thought the origin of the name Imbolc came from the Old Irish *i mbolc*, meaning 'in the belly', referencing the symbiotic relationship between the ancient people and their livestock. Is it better to spend as much of our year as possible in warmth and sunlight? Or should we stay in our natural environments, absorb the rhythm of the seasons and find the joy in every breath of winter wind, every broken leaf and frozen raindrop?

Each year there is a day between February and April when spring reaches out from the cold depths of winter and taps you on the shoulder. It is an awakening. Perhaps the weather reporter has called it 'the warmest day of the year so far', or perhaps it is just cloudless, windless, the scent of new life on the air. You feel warmth on your face after so many insulated months of half-light, half-living, tucked away inside to escape the claws of winter. An inch of skin exposed to the air is brushed by spring; the sky seems more blue, more alive, more merciful. And deep within your stomach something stirs, a longing for the outdoors that has been hibernating like a great bear, otherwise content to be wrapped up inside with tea and telly. The bear in you wakes, restless, stretches out his thunderous paws, shakes away the cobwebs of January, looks up to the sun and remembers long, hot days filled with wild roses, river swimming, cider in pub gardens, fresh raspberries, sunburn, bare feet, salty chips. Summer is on its way.

Once the burn had finished, most of the congregation started slowly moving towards the gates, back to their cars and the comforts of modern life. I left the space in which I'd been sitting and moved away from the crowd, back towards the roundhouses and Roman villa, where a few exhibitors were packing away their things and it was emptier, more peaceful. Under the light of the stars I walked along the border of the farm where everything had been left to grow wild. The flowers had closed their petals until morning when the first light of spring would warm them from root to tip and open up their colours to the dawn. And while most of the birds had now presumably gone to sleep, one blackbird was still singing beyond the trees, its rich, velvety song serenading the crowds, calling farewell to the singed wickerman and welcoming in the warm summer months ahead.

In the darkness everything smells different – more damp, more fragrant, more earthy – and as I stood alone in the corner of the farm, the air was now swelling with new life and woodsmoke, my favourite combination. In the ditch around the Iron Age enclosure, I could see new shoots of rosebay willowherb arriving from the depths of the soil, and by summer I knew the entire bank would be thriving with hot-pink flowers. It's a pioneering plant, the first to colonise barren lands with very little vegetation; in Britain it is also known as 'bombweed' as it was one of the first plants to recolonise bomb craters after the Blitz in the Second World War. In winter it shrivels right back to nothing, but by July it would be a shimmering cascade of colour swaying in the wind, full of bees and butterflies, a perfect pink enclosure for our beautiful Iron Age village.

I wandered over to the Roman villa, pale in the twilight with a grape vine growing up the wall and neat pots of herbs along the front edge. In the garden the walnut tree was full of new leaves, rich and shining, and if you looked closely you could see the humble beginnings of new buds coming through, tiny green pods that would later develop into crunchy walnuts, ripe on the tree. We only ever had a few nuts growing each year but it was always a treat; we would share out the spoils between the office staff, one small section of walnut brain in each eager hand.

Inside the kitchen of the Roman villa – or at least, what was presumed to be the kitchen, as the only archaeological evidence from the original site was a hearth – a statue of the Roman goddess Ceres had been placed in an alcove in the wall. She was the goddess of agriculture and fertility, and a Roman household might have left food by the statue as an offering, encouraging Ceres to bless them with a fruitful harvest. No full-time Roman family lived in our villa, so instead she was surrounded by dried flowers and paintings, our own salute to the Roman gods and their mystical powers. I walked into the villa, which was now empty of visitors who had filtered down to the wickerman paddock, and found Ceres sitting there in the dark, a fresh cobweb woven over her head.

Now and then, I wish I could believe in a god, so as to feel like someone out there has the power and benevolence to give my life a little boost or protect me from harm. The Romans had hundreds of gods – one for every purpose and every problem. Need to win a battle? Call on Mars, the god of war. Hoping for a successful marriage? Make an offering to Juno, the goddess of matrimony. Need to revive a withered apple tree? Get hold of Pomona, the goddess of orchards. Worried about that volcano erupting? Vulcan's your man – the god of fire. Many of the Roman gods came with a Greek equivalent, such as Venus and Aphrodite, because when the Romans ruled Greece they adopted the Greek belief system, gave them new Roman names and changed some of their mythologies to better suit their own belief systems. In Greek mythology, for example, Zeus frequently visited earth in disguise and got up to all kinds of mischief, starting fights and impregnating mortals, whereas the Roman god Jupiter ruled from the heavens, never debasing himself by coming down to earth and mingling with the plebs.

The Greeks worshipped several deities associated with the darkness, starting with Nyx, the goddess of night. Held in great esteem by the ancient people, she was believed to be one of the oldest primordial gods born from the void of Chaos, together with Gaia (Earth), Tartarus (the Underworld), Eros (Love) and Erebus (Darkness). Nyx became the mother of many other figures in the Greek pantheon, but she was usually depicted as a woman holding two children in her arms: Thanatos, the god of death, and Hypnos, the god of sleep. She also gave birth to the Oneiroi, a group of gods and demigods that ruled over dreams and nightmares, fathered by the god of darkness, Erebus. The most famous of these were Morpheus, who took the form of men, Phobetor, who took the form of beasts, and Phantasos, who appeared as inanimate objects. The next time you suffer a nightmare, remember that it's probably just Morpheus, Phobetor and Phantasos on a lads' night out.

The Greek and Roman gods are surrounded by so many stories and symbols, they still capture our imaginations today. One of my favourite computer games growing up was Disney's *Hercules*, which was based on the Greek legend of Heracles and took my stepsister Christie and I months to complete. A Trojan Horse is the name now given to a virus that is allowed into a computer by disguising itself as something benign, based on the Greek tale of the Trojan War. Sigmund Freud took inspiration from Oedipus to name one of his most famous psychoanalytic complexes, and even the multinational sportswear corporation Nike is named after the Greek goddess of victory. The names of constellations are also well known for their Greek and Roman origins, and the US Apollo Space Programme, carrying astronauts to the moon, was named after the god of the sun and knowledge who was known for his skill as an archer to always hit his target. One of the most fascinating ecological theories of the twentieth century – the Gaia Hypothesis – was named after the mother of the Greek gods by its creator James Lovelock, who argued that all organisms on earth interact with their surroundings to form a self-regulating system, which maintains the conditions for living things to survive on the planet. The Greeks and Romans, and most of the ancient civilisations, were intelligent and culturally creative, so it makes sense that we have held onto their ideas for thousands of years.

There in the darkness of the kitchen, I took a closer look at the statue of the fertility goddess Ceres, faintly illuminated by the moonlight shining in through the windows, then cleared away the cobweb and rearranged the flowers. By the time I returned to the beer tent, where I had spent most of the evening volunteering with friends and family, providing pints to happy festival goers and pouring pale ale over myself in the process, most of the crowd had started filtering out through the exit. As the farm only has one small car park, each year the farmers allowed us to use one of their fields to accommodate everyone's cars, which meant that at the end of the festival guests had to take a short walk up the hill to find their vehicles and start the journey home. To help guide them, we gave out small flaming torches (safer than they sound), so that to watch the visitors walking up the hill was like watching a line of explorers off on an adventure into the night, the silhouettes of families black against an empty twilight sky, perfectly blue and clear. It somehow reminded me of the school nativity play – perhaps the wise men following the star over the hill with their torches ablaze.

I thought of our ancestors celebrating this festival in the past, and all the festivals that marked a change in season, a shortening or lengthening of the days, the time to sow or harvest, to be happy with the warm weather or to prepare for the long, cold winter. How much did these ancient rituals mean to them? If they did celebrate Beltain in this way, did they genuinely believe that by burning the wickerman the gods would bring them a successful harvest? Or was it, just like today, a simple excuse for merriment, for happiness shared in the warmth of a late-spring evening? Aside from the visitors who had come to the farm this evening, how far had we distanced ourselves from these seasonal celebrations, and did we have any rituals left at all?

Perhaps not in the official sense, but in Britain we are still obsessed with the changing seasons and weather. For me, summer is not so much associated with farming or preparing for harvest, but with spending long, warm evenings in the Queen's Head pub garden. It's not the only way I spend my time, but in the depths of winter when I am cold and miserable, sick of the greyness that cloaks the countryside and the stagnation of the landscape, I imagine the pub garden as my happy place in summer: skin exposed to the sun, cold cider in pint glasses, hot pizza, and grass between my toes, surrounded by all the friends that have grown up in our hometown, who may have all gone on to different careers and lifestyles but who understand the beauty of being there together, the unique joy of sitting in that garden under the evening sun.

One night we had been lounging around in the pub garden for hours, drinking and laughing and bathing in the heat of the evening. There was me, Dave, his brother Andy and the family Schnauzer, Tinks, and we had been sitting outside so long that we were all baked. We decided to leave the pub and head down to the river on the other side of the town, a quiet stretch of the River Rother that weaves from Empshott in East Hampshire, all the way through our town and over to Stopham in West Sussex, after which it joins the River Arun and flows out towards the sea. The upper half of the river was used for centuries to power watermills, first recorded in the Domesday Book in 1086, but today it has less industrial associations and has been officially recognised as a Site of Nature Conservation Importance for its valuable wildlife habitats. In 2001, a camera trap picked up a sighting of an otter – the first confirmed evidence of otters returning to the Upper Rother in 14 years. The presence of otters is a natural sign that a river is healthy and, since their reappearance, conservationists have worked to improve the water quality and to reduce pollution running off the surrounding land, in the hope of keeping the otters there and increasing their population.

We arrived at the river and started walking down to the water's edge. By now the sun had almost set, and a golden light filtered across the sky between blushing waves of blue and pink. It was the height of summer and everything was wild with growth; this place marked the start of a walking trail called the Serpentine Way, but it was so out of the way that it was free from the usual restrictions of highways and byways, and the hedgerows had been left to grow long and straggling rather than being forced into shape by a hired man and his hedge trimmers. The grass was tall and bristling, full of birds and mice and the insects I tend to avoid unless wearing long trousers.

We wandered down to the river, Tinks lolloping about with her ears bouncing, her tousled face full of joy at being free to explore after being well behaved at the pub for so long. We walked through a tunnel of trees, over a fallen log, rotten and smothered in lichen, and into the welcome shade of the forest in the evening heat. There the river opened up beneath the tree canopy, a perfect stretch for entering the water and swimming across the riverbed. The bank rolled down into the water and we could walk across the soft earth with bare feet.

Always prioritising efficiency, I had been wearing my swimming costume under my clothes and so I quickly undressed before edging closer to the water, reluctant to submit my baking skin to the cold river. I tiptoed in, the riverbed full of shingle that was not painful to touch but still worth taking a bit of time over each step. One, two, three. Into the water I went, up to my thighs, water surging around my waist and bringing bumps of shock to the surface of my skin. The water was so shallow that I spent a while only waist-deep, enjoying the heat of the sun on my shoulders and the water around my legs. I looked back and saw Dave and Andy still making their way in, and I could read the same dilemma on their faces: the eternal debate between being boiling hot and succumbing to the raw chill of the river water. Eventually they waded in, so that the only one of us left on the bank was Tinks, looking beautiful and shining in the rays of the dying sun.

It became clear that Tinks was even more wary of the water than the rest of her pack, and I forgot how different dogs were suited to different habitats. Our old golden retriever Murphy loved the water, flumping into the local lake to chase the swans and float through the blue with his eyes shining. Afterwards he would emerge onto the shore like a swamp monster, his fur shrunken into soggy strands until he shook his whole body, freeing and fluffing himself while drenching us all in wild water. As a golden retriever, with his oily coat and webbed paws, he was bred to love swimming. But Tinks was smaller and more dainty than Murphy, and it took her a while longer to get used to the idea of coming in. In fact, if we hadn't started swimming up river and away from where she stood stranded on the shore, she probably would have stayed on the bank the entire time. As it was, she became more anxious as we moved further and further away, so in the end we waded back over and helped coax her in. It turned out she wasn't a fantastic swimmer and, although she was able to float, she was so small that it took all her energy not to sink, so instead we took it in turns to pass her between us in the water like teaching a child how to walk. One of us would release her gently into the river, and the other would call her over with cheers and smiles, until she was safely back in our arms. Not the most efficient way of moving through the water but after a few goes she enjoyed herself, and in our strange triangular position we spent the next hour or so paddling about in the water, our skin cooled off from the heat of the sun, the trees trailing their leaves through the surface of the river and softly over our shoulders. All the time the sun was setting, the light failing, the landscape cast into peace and shadow.

By the time we finished, refreshed and damp, it was dark, and we wandered back through the fields, through clouds of midges buzzing lazily in the twilight, through the trees with sleeping birds and waking owls, through the dirt path beneath a darkening sky speckled with stars. On a clear night we could always see the stars here, piercingly bright against the darkness, constellations in every direction, telling the stories of gods and goddesses, warriors and beasts. These were the same stars that had been observed for thousands of years by human civilisations, stretching out over our heads in their glittering array, older than the earth itself.

TIF

CHAPTER NINE

Midnight Sun

There was a kestrel hovering on our runway, floating over the grass verge between the taxi point and the flight path, seemingly oblivious to the 300 tonnes of aluminium and carbon fibre hurtling past every few minutes. I was sitting on a Boeing 737 at Gatwick airport, and our plane was currently in a queue of several other aircraft waiting their turn to heave onto the runway and leap off into the sky. I'd already eaten all my Starburst and my phone was switched to aeroplane mode so, with nothing better to do, I waited and watched the kestrel floating like a drone over a bleak patch of grass in the middle of the concrete.

Kestrels are the only birds of prey in Britain to hover in mid-air, although buzzards sometimes manage to catch the wind current and float in one place, wings folded right back into an 'M' shape. For those who enjoy a little motorway birding (my most likely cause of death), it is the kestrel that is usually seen on the edge of the road, wings beating, unmoving while it lingers in one spot over a rodent on the ground, waiting for the perfect moment to dive down and snatch it. It's yet another example of how nature has perfected the creation of birds, their aerodynamic body shapes inspiring many of the man-made flying machines we have attempted to build for ourselves over the last few centuries.

I watched the kestrel dangle over the grass, uninterested in the gigantic metal beasts rolling by that looked vaguely like other birds. It floated, impossibly still like a paused video screen, before suddenly tumbling from the sky to capture its prey. Before I could watch it emerge again, our plane started to move, and I settled into my seat to prepare for my next

1,600km journey across Denmark and Sweden, over to the vast forests and glistening lakes of southern Finland.

It had been six months since I travelled alone to Tromsø in Arctic Norway. Six months since I watched the aurora borealis dance across the sky, eider ducks huddled on the water beneath the golden glow of the *Ishavskatedralen*. Now I was flying to the Finnish capital of Helsinki, renewed and at peace with the world that was disappearing below me as we rose higher into the clouds. I suppose the best word to describe the last year of my life would be 'intense', but even now the cloud of doubt and misery that had swept over the previous autumn and winter was vanishing from memory. I felt like a moth emerged from its chrysalis; life as a caterpillar had been sweet enough, but now I could see everything with new eyes, a veil had been lifted.

I'll be the first to admit I have a little obsession with Scandinavian culture. Geographically Finland is not actually part of Scandinavia, and culturally they haven't made their mind up either. Half the population claim they are advocates of the Scandi lifestyle, but the other half say it is the Nordic culture they are most associated with. I'm not bothered about the label; it's the ethos behind the societies of Norway, Denmark, Sweden, Iceland and Finland that I'm most drawn to, all of which rate consistently high in the UN's annual World Happiness Report. In Britain, it's fair to say we are in the midst of a Scandi cultural takeover, but there's more to their happiness than beards, minimalism and lots of coffee. Our bookshops are filled with little hardback guides to Nordic living, from the cosy comfort of Danish *hygge* to the art of balanced living with Swedish *lagom*. They may just be stocking-filler books with beautiful illustrations, but they give us a glimpse into a more peaceful way of living that many of us might neglect.

In Finland it's all about *sisu* – the ancient art of courage, resilience and grit in the face of adversity. According to *sisu*, the challenges of life should be faced thus: personal wellbeing should take priority; communication between friends and family should be clear and effective; a healthy mind and body should be maintained; children should be raised as kind and resilient little people; and (my favourite part of all) we should fight for what we believe in to make the world a better place. In 1940 *Time* magazine called it 'the ability to keep fighting after most people would have quit', and it was this that first drew me to perhaps the fiercest of the Nordic countries, and one that is seriously engaged with social and environmental issues.

Aside from its Nordic charm, however, I soon realised that Finland has an identity entirely of its own. Every sign was written in both Finnish and Swedish, and I was amazed at how different they sound. Having listened to my sister and her friends speak Norwegian, I ignorantly assumed Finnish would sound similar, but in reality it seems to have more of a Russian twang, despite Finnish and Russian being from different language families. It still sounds beautiful (as most languages do to those who don't speak them), but the bouncing softness of Norwegian and Swedish is replaced by a harder edge that seems to resonate with the complexities of Finland's Soviet history.

I arrived in Helsinki and found my way to the apartment I had hired in the city suburbs, where I dropped my things, made a quick coffee and headed back out into the hot, bright weather of the Finnish midsummer. This was the reason I had travelled here for a long weekend in June. Every year across the Nordic countries, even as far south as Helsinki, the summer months are illuminated by the midnight sun, a natural phenomenon where the sun barely sets and the sky never goes fully dark. In Tromsø, where I had been immersed in the polar night just six months before, the sun would remain in the sky from May to July, and the sky would now be drenched in the blood-orange tint of eternal sunset. Knut Hamsun wrote in his novel *Pan* how it was like the sun had dropped into the sea for a drink, and returned again, refreshed. Here in Helsinki, the sun would at least disappear beneath the horizon, but by such a small amount that even at one o'clock in the morning the sky would be alight with the first greyness of dawn, the birds already awake.

The heat of the morning had quickly climbed to 23°C, so I decided to embrace the midsummer weather by exploring the shaded forests outside the city. Two bus journeys and one train ride later, I reached Nuuksio National Park, a 53km^2 collection of lakes and woods spread out across the regions of Espoo, Kirkkonummi and Vihti. I had heard the forests here were bursting with wildlife and I was desperate to explore, inspired by the ancient Finnish hikers' code of *jokamiehenoikeus* (literally 'every man's right') that offers everyone the right to roam through public and private spaces as long as the land is respected. It took me a while to resist the idea that I might be trespassing on people's land, but with the encouraging smiles of the one or two other hikers I encountered along the way, I soon relaxed into a state of exploration, liberated from the constraints of fences and signposts, free to discover new spaces and hidden slices of wilderness. At two different points I got lost; in a country covered by 75 per cent forest (in Britain it is currently 12 per cent), when you're not following a path it is remarkably easy to become totally disorientated, lose phone signal, get eaten by giant ants and stumble knee-deep into what *looks* like a dried up riverbed, all before remembering that Finland is also home to wolves, bears, lynx, wild boar and racoon dogs.

I survived nonetheless, walking for 16km through coniferous groves, lakesides and well-worn dirt paths to find hazel grouse, wrynecks, three-toed woodpeckers and a pile of elk droppings (although sadly, no elk). I spent most of the day scouring the forest floor in search of yellow and orange rice-shaped droppings, hoping they would lead me to the nesting spots of Nuuksio's mascot species, which was being successfully conserved here with help from the park team: the Siberian flying squirrel. Sadly I didn't find one – they are famously elusive and primarily nocturnal – but they are one of the cutest species I'm aware of. The park was full of illustrated signboards displaying their fuzzy grey faces, large black eyes and tiny paws, recognisable for the patagium membrane that stretches from wrist to ankle and allows them to glide silently through the trees.

By eight o'clock in the evening I was full of that glorious exhaustion that comes from a cocktail of fresh air, sunshine and sweat. I reached the visitor-centre shop just before it closed, and was now sipping a cold bilberry juice by the side of the lake, watching a pair of fieldfares rustle around in the shrubs at the water's edge. Like redwings, fieldfares are also winter visitors to Britain, where they migrate over from colder regions to find food and shelter in the milder climate of the UK. By spring, most fly away and return to their home countries of Norway, Sweden, Finland and Russia. I was now watching them in their summer habitat and it was wonderful, like interrupting them on their summer holiday. With mottled plumage of greys and browns, they are beautiful birds all year round, but there was something about watching them play in the sunshine that brought them to life. For a bird that is usually seen against the drab, frostbitten backdrop of the British winter, here the reflected light of sun on silver birch shone down on their feathers and seemed to revitalise them, and it was easy to imagine that they too were celebrating the height of midsummer, the warmth and magic of the endless midnight sun.

With a purple tongue I finished my juice and checked the Helsinki transport app on my phone. By my calculations, I had two hours to reach the bus stop and jump on the last bus back to Espoo, where I could then take the all-night train back to Helsinki. Plenty of time to walk back through the forest and up to the main road. I hoisted my rucksack, double-checked the route and started walking, turning to the lake one last time before I disappeared into the trees. Eight o'clock was barely early evening in midsummer Finland, but something had changed in the air, and despite the usual brightness of the sun I could tell it had started its long, slow descent beyond the horizon. It would be several hours until the sky grew vaguely dark, but already it had shifted from white-blue to the palest lilac, marking the beginning of the short night.

I continued walking towards the flashing beacon on my phone, looking around every few seconds in the hope of finding the flying squirrels that I was certain would float down in front of my face like an autumn leaf. No squirrels, but after a few minutes of walking I heard something else instead, a sound that I half-hoped might belong to another species I had wanted to find on my journey into these woods. Earlier I had followed the unmistakable drumming of a woodpecker, leading me up a steep slope to a grove in which I found a three-toed woodpecker prancing about on a birch tree, hammering at the wood to announce its presence to the rest of the forest. It looked just like one of our spotted ones at first, but then I noticed a faded yellow smudge on the crown of its head, a darker wing and much denser spots along the back. I watched it linger on the trunk of the tree, hammering every few minutes but otherwise gazing around to see if its fellow woodland creatures were suitably intimidated, until I retreated carefully back down the slope to the path. I had learned my lesson, and did not want to get lost a third time or be eaten by a large, toothed mammal.

Now I could hear the same sound again – that incessant hammering reverberating through the trees – but this time it was different. The three-toed woodpecker was small, and consequently its drumming was higher pitched and tinny. This was another bird, and I was almost certain I knew which one. The drumming was deeper, heavier, echoing through the forest and into my heart muscle in two-second bursts, like when we used to flick a ruler on the edge of the desk at school and wait for it to stop vibrating. But trying to find a woodpecker is difficult for a number of reasons. For one, they only drum in short bursts so most of the time you are following silence; for another, they usually sit perfectly still, so you have to hope they have settled low down on the nearest side of the tree, or you'll be trapped in a maddening vortex of drumming without ever finding the source.

I picked up my pace and started following the sound through the trees, trying to avoid the crunchiest mounds of leaves and dead moss so that my presence might go undetected. Through towering birch trunks and rotting stumps, past hidden lagoons and granite rocks I climbed further into the forest, following the woodpecker hammering on the wind, until at last, after a long hike away from the dirt path, I reached a valley of conifers and silver birch trees so tall that only the crown of the canopy could reach sunlight, bathed in the soft peach tones of the sinking sun. In the time I had been walking, the sun had continued to drop lower and lower in the sky, so in the gaps between the trees I could now see a pearlescent mauve fading into blush, spreading out across the heavens. Below this, the trees were already draped in shadow, and in a cluster of lower branches a gang of long-tailed tits dangled like toffee apples.

This was where the woodpecker had taken me, and in the dying light beneath the trees I could still hear, every now and then, the same deep drumming echoing through the valley. The forest acted like an amphitheatre, and it took a long time to gauge where the sound might be coming from, tilting my head back and forth like a barn owl. Eventually I followed it up one side of the valley, clambering over rocks and roots, not daring to stop for more than five seconds lest the bastard wood ants bit me again. Up and up the hill until finally I came to the summit, where a clearing had formed in the trees and the floor was layered with granite slabs and dead moss. Here the drumming was so close that I could feel it in my head, along my bones and echoing out through the pores of my skin. Where was it? I knew it was within reaching distance, but the trees were bare. I moved towards a large conifer and, in doing so, stepped on a thick swathe of dried moss, which crunched deliciously underfoot and broke the peace of the evening air – and there!

In my clumsiness I had finally spooked the woodpecker out of his hiding place, and in the clearing, just feet from where I stood, appeared a bird that was both so beautiful and so strikingly menacing that I could only utter 'Oh' and take a small step backwards.

I knew that black woodpeckers were frequently cast as the villains of European folk tales, but until now I didn't know why. In front of me was a space encircled by trees, and through that space now glided a spectre: a ghoulish, shining beast with a long, ivory bill and piercing yellow eyes like pools of liquid gold. The feathers were black, shimmering in the evening light, and on the crown of its head a crimson tuft shone bright against the shadows of the trees, a red flag of passion, of terror. But it wasn't the feathers or the eyes that made me recoil, nor the tuft like a smear of blood. Unlike the wavering, bounding flight of the woodpeckers at home, laughing across the fields and farmlands like a hooligan, this giant bird was almost motionless in flight, gliding through the grove with wings outstretched, head raised, eyes focused. It was perhaps the head that caused me to feel so uncomfortable; most birds fly with their heads down, streamlined and smooth, but this creature looked twisted, contorted into a pose like a waxwork figure. I was reminded of a Nazgûl, those foul wraiths from *The Lord of the Rings* that, when unhorsed at the Ford of Bruinen, continue to haunt Frodo and Samwise on their winged, serpentine fellbeasts.

Perhaps if I'd had time to observe the woodpecker more closely, the horror would have faded and I would have memorised the beauty of its glossy feathers, the strength of its bill and the intelligence captured in the curve of its eye. But my time with the bird was momentary; within seconds it had swept across the forest in front of me, eyes fixed on the trees ahead, moving but motionless in flight. As fast as it arrived, the woodpecker disappeared into the darkness and, after a few seconds of silence, that bewitching drum started beating through the air once more, the voice of a haunted spirit fixed to its birch totem.

Stepping over a pile of elk droppings, I climbed back through the valley and down to the dirt path. The night had drawn in quickly, and through the trees I could now see the sun touching the horizon, a great golden light oozing through the forest, casting everything into a bronze glow. The sunset was slow here, an everlasting haze of warm light dancing off the leaves, the earth, the rippling lakes. Back home, the sunset would signal the end of the day, a time for rest and peace, but here it felt different; the forest was drawing me back in, away from the busy beauty of the city and into the nocturnal world of a woodland lit up by the half-light of a sky that never truly darkens. Why go home when the forest never sleeps?

I had been chasing the woodpecker for longer than I thought, and it didn't take much to realise that even if I reached the road in time, I had now missed the last bus back. Not the greatest feeling if I were in the woods back home, let alone a strange country with more trees than rollmops. I resisted panicking – such a waste of time – and instead accepted that I could do nothing but start to walk the three-and-a-half-hour journey back to the train station in Espoo, from where I could get an all-night train back to Helsinki and home. A ridiculous plan, but I refused to pay for an extortionate taxi and thought I might be able to get a lift on the way, so I started the long trek home with the most cheerful disposition I could manage and a slight hunger in my belly – a celebration of *sisu* if ever there was one. At least it wouldn't be dark for my journey, and for 20 minutes

I walked along the road under a lilac sky, listening out for owls or bats, swatting away the odd mosquito that dared to come close. And then, like a saviour in the night, along came the first car I'd encountered, and I did what they do in the cartoons and waved it down. The driver was a very Finnish-looking man called Colin, around my age and quite good-looking (although with a hint of Nazi Rolfe from *The Sound of Music*), and he offered to drive me all the way back to the train station.

He asked me about England and told me he had only been there once, to Birmingham for a team-building weekend with the construction company he worked for in the summer. In winter he ran a ski slope outside Helsinki; we had to drive there first to unload some equipment, and the place looked bleak without snow or skiers. I told him about my day in the forest and how I was disappointed not to find any flying squirrels, but he pointed out that he had lived here his whole life without seeing one, which was some consolation. We talked about the forest and how lovely Finland was, and he told me how everybody migrated to Helsinki from the north, which was a shame because the rest of the country was so beautiful but empty of people who could support the local communities there. I told him about my interest in the midnight sun and he indicated it was a season of joy and exuberance; he and his friends would be going fishing later that night long into the early hours of the next morning.

When I finally arrived back in Helsinki city centre it was nearing 2am, and the streets were filled with Saturday-nighters flitting between bars, dancing in the street with bottles in hand, the strange half-light illuminating their faces. The city was alive, as though every resident was trying to forget the long, dark winter by soaking up every drop of sunshine released into the sky, desperate not to miss one moment of it. It was a wonderful atmosphere, and if I hadn't hiked 16km, been bitten to death and been wearing hiking boots, I would have been tempted to immerse myself in it. As it was, I was utterly exhausted and dreaming of a hot shower, but before nearing the bus station for the final leg of my journey home, there was still one thing I needed to do. A few months previously, I'd read about the new McVegan burger being trialled in Finland, and I couldn't waste the opportunity to indulge the more depraved, heart-attack-inducing side of my taste buds. I found a 24-hour McDonalds, loaded up on chips and scoffed my burger on the bus back home. It was, unfortunately, thoroughly delicious, and as I forced chips and Coke into my mouth like a sleepy pig, my mind floated back to the woodpecker drumming away in Nuuksio, miles from the noise of the city, echoing through a silent forest where fieldfares danced in the sunlight and squirrels drifted through the air like ethereal feathers.

I rolled out of the bus at around 3am, and started the slow walk back through the park, along the road and over to my apartment. The air was sweet, and I glanced around in confusion at where the aroma was coming from, before realising the roadsides were planted with swathes of wild rose bushes bursting with bright-pink flowers. The early-morning light had already encouraged their petals to spread, and from the centre oozed that delicate fragrance that makes the world smell like Turkish delight. I breathed in deeply as I walked, filling my lungs and brain with the scent, displacing the artificial horridness of my now-digesting McVegan burger and chips.

The park was a circular, grassy space encircled by wildflowers, with an enormous mound in the centre that provided pleasant views of the fairly flat countryside on the outskirts of Helsinki. Somewhere in the distance I heard a horse neighing, and the birds continued to sing in the trees. Suddenly, I stopped. I had been gazing at a large rabbit on the other side of the park, sitting round and squat in the flowers, nibbling on a stem. I love rabbits, and remarked to myself how big this one was – and what long ears! And then

I looked closer and realised it wasn't a rabbit at all. I was so used to being in Britain, where a small mammal is likely to be a rabbit, that I didn't even think it might be anything else – let alone a hare.

A hare! My only real experience of a hare had been a fleeting glimpse down a green track one afternoon at home, and as soon as it had clocked me it was gone, vanishing into the hedgerow in a tangle of muscle. In Britain they are elusive creatures, not impossible to spot but certainly less observable than the rabbits lingering on motorway verges and flumping about allotments in search of greenery to loot. But to see a hare so openly here, in this suburban park on the outskirts of Helsinki, was so surreal that, combined with my aching legs and exhausted brain, I was overwhelmed with bewitchment.

Just seconds after spotting the first one, I saw another come bounding over, a lolloping tangle of black-tipped ears, diamond eyes and velveteen feet. Soon they were joined by two more and, despite my tiredness, I couldn't resist sinking into the grass to lie down on my front and watch them dance in the pale light of early morning. The first stayed in the wildflower verge, chomping on the long grass that concealed its body so that only the large eyes and ears stood out above the green. Two more spent their time racing around the park, leaping, play-fighting, squabbling over some conflict I would never understand. The fourth seemed more relaxed, languidly hopping across the grass with its long paws and hind legs almost dragging behind.

Many of the folk tales in Britain about rabbits are in fact about the native hare, as rabbits were only introduced from Europe between the Roman and Norman periods. The Easter bunny itself is thought to be a hare – the consort to Eostre, the Anglo-Saxon goddess of spring – although there is very little historical evidence for this connection. The Iceni queen Boudicca famously kept a hare inside her tunic before a battle, and when she set it free the hare's path was interpreted as an omen for their chances of victory. Due to their strong legs and sudden, powerful leaps, the hare has also been associated with dawn, new beginnings, rebirth, the lunar cycle, women and fertility. They were so closely associated with fertility, in fact, that it was Aristotle who first suggested they might be able to get pregnant while pregnant (something that has now been proven with modern science: a male can fertilise a female during the end of her pregnancy, but the embryos will only develop a few days before she is due to give birth, ready to move into the uterus after the first litter is born).

For a while longer I watched the four hares tumbling around the park until at last, overwhelmed by sleep, I wandered back to the apartment for a hot shower and bed. The next day I woke late and lounged about until noon, recovering from the previous day's expedition, before heading out to explore the vibrant city-centre streets of Helsinki. I found a flea market near the water's edge selling all kinds of beautiful homeware, Finnish ceramics, Moomin paraphernalia and clothing. Fortunately for my bank balance, I had only brought an already bulging hand-luggage case, so I would be unable to buy everything my heart desired. In the end I settled for a sage-green enamel kettle, a retro ski jacket and a striped rug, all of which I somehow managed to cram into my suitcase later.

The people here were in full summer swing, and a sense of joy radiated from everybody I spoke to. They seemed to nurture a stronger connection with nature than most other countries I had visited, combined with a proud environmentalism that meant the streets were clean and the public transport was excellent. Aside from the glitz of the central shopping centre, the minimalist essence of Nordic culture had resulted in a clutter-free existence away from consumerism that seemed to celebrate the bare essentials of happy living: good food, friendly conversation and lots of time spent outdoors, especially in summer when the nights were so short and the landscape was bursting with life and beauty. In Nuuksio the information boards warned against the usual things – don't start fires, don't get lost, don't drown – but there were also large colourful panels dedicated to the things you *could* do. Foraging for wild food was encouraged, which I knew was a common part of Finnish culture, and the board pointed out a few plants and mushrooms that were good to eat in the depths of the forest. Dogs were welcome but required to be on leads to protect wildlife, particularly birds who build fragile nests on the ground that can so easily be trampled by dogs. Our wild spaces are for everyone to enjoy, but there was a sense of respect here, of living in harmony with the outdoors rather than treating it like an asset to serve our needs alone.

By evening I had wandered back to the apartment to spend an hour or two relaxing indoors, catching up on emails and listening to the swifts pouring through the air outside. I swore they were even louder here than at home, but perhaps this was because I was on the fourth floor and more in line with their acrobatic displays. I opened the windows wide and watched as the birds cascaded through the sky, spinning and looping like trapeze artists, all the while sending out those fierce screams that had come to symbolise the long, hot days of summer at home, lolling about in the garden with an iced gin and cucumber. I felt sleepy, but the weather was still so beautiful outside that it felt wrong to be in the apartment, even though the idea of walking another kilometre might finish me off. So instead of exploring any more of the land, I decided to take the bus down to the shore and spend time with the water.

It was only a half-hour journey to the coast, and ever since reading *The Summer Book* by the Finnish writer Tove Jansson, my mind had been swimming with archipelago landscapes, windswept islands in cerulean waters, driftwood washing up against dirt and sand. Jansson – one of my favourite women of all time – is perhaps better known for her children's books than her adult fiction. Most of all, she is known for her creation of the Moomins, a family of white, rotund creatures with large snouts who look like hippopotamuses and live a life of carefree adventure in their house in Moominvalley. I am unashamedly obsessed with them, not just for the beauty and sweet simplicity of their stories and illustrations, but for everything they represent: a life of love, joy and adventure lived outdoors in nature, epitomising all that is good and important in the world. As an author and painter, Jansson has long been a growing source of inspiration to me, and part of my excitement in visiting Finland was to experience for myself the landscapes that fuelled so much of her work. In her novel *The Summer Book* (or *Sommarboken*), she writes about an elderly woman and her granddaughter who spend one summer together on a small island in the Gulf of Finland, near to where Jansson herself grew up.

The archipelago of Helsinki consists of around 330 islands, from sea fortresses and Viking defence points to foraging hotspots and nature reserves, while the city itself is known as the 'Daughter of the Baltic Sea'. Many of the islands have become pleasant tourist destinations, particularly around midsummer when visitors can enjoy boat cruises, bonfires and endless parties under the glow of what they call the White Night. My bus took me down to Marjaniemi, just east of the city centre and close to the island of Iso Koivusaari. Unsurprisingly, there were a few other people here too, enjoying the evening weather and low sun streaming across the water. On a large rock jutting out to sea, a group of teenagers were drinking and laughing, and a few metres down a young couple were canoodling on a stone wall while the odd cyclist rolled past on a night-time ride in the twilight.

I walked further down the beach, past a father and daughter playing in a rockpool beside a group of snoozing barnacle geese, until I came to a wooden platform sticking out into the sea and a flight of stairs descending straight into the water. Across the sea there was a tiny island, inhabited only by tall trees and one small wooden cabin painted crimson red with cream windows. I wondered who lived there, and whether it might feel isolated to be on an island all alone; then again, the boat trip would probably take just a few minutes – and what a peaceful place to retreat to, leaving behind the noise of the city with nothing but the wind in the trees and the sea foam lapping against the shore beneath an endless sky.

The water was calm, its waves diffused by the islands so that the surface was a sheet of ripples moving with a slow tide. I noticed two swimmers floating along a line of red buoys, and speculated how cold the water might be. My only experience of Nordic temperatures had been the melted glacier water in the Norwegian fjords where we went white-water rafting, and in which we had to first complete a swimming test to prove we could cope if we fell out – which I did. Even in wetsuits the water in Norway had been bone-breakingly cold and, although the sun now shone encouragingly on my warm skin, I resisted taking the swimming costume out of my bag until one of the swimmers approached the shore and climbed up onto the stairs.

'It's very warm!' he said with a friendly gesture, and I returned his smile with an unconvincing 'Good!' while guessing the Finnish idea of 'warm' might be slightly different to mine. But he did have kind eyes, and as he headed off home I decided to risk it. I love swimming outdoors, and although I am always afraid of the first hit of ice-cold water on my skin, I know how quickly the body numbs and adapts. Besides, I couldn't imagine a more beautiful spot to finish my weekend, to clean away the ant bites of Nuuksio, the heat of the city centre and the sunburn on my shoulders.

I changed into my costume and left my bag on the side, not at all worried that someone might take it; Finland has one of the lowest crime rates in the world. As I approached the stairs I could see the other swimmer, a woman with short red hair, gliding away at the furthest point of the beach. I descended the first steps until I could feel water lapping at my toes, the soft squelch of algae slipping over the soles of my feet. One step, two step, and to my surprise the man had been right. Perhaps it was because the sunlight had been shining on the water for weeks or perhaps it was simply because the outside temperatures were so high, but it genuinely felt warm – far warmer than anywhere I'd been swimming in Britain that year. My usual cowardice at the slightest droplet of cold water melted away, and I wandered into the sea with a boldness quite unlike my usual self, until I was shoulder-deep in swirling blue waters under a majestic sky.

It was a strange and beautiful experience; I was used to either swimming in rivers – fresh and wild but enclosed by the bank – or in the sea where the waves come at you like a slap in the face. In the archipelago it was different. I had expected salt water, but in truth it was neither fresh nor salty, instead an unfamiliar metallic flavour that was not unpleasant. And although there was a current moving through the water, revealed by tiny ripples like contour lines on a map, it was so gentle and slow that to swim against it was effortless. I allowed myself to bob up and down and watch how far the sea would take me. Perhaps the most joyful part of swimming here was the boundary – or lack of it. I was literally at sea, protected from the tempestuous Baltic only by a cluster of islands, but their strength meant that I could swim about in freedom, limited only by the horizon into which the sun was now sinking. There were no riverbanks, no flags warning of strong currents; just me and the water, floating around in the ocean with nothing but blue as far as I could see and oystercatchers crying out over my head.

I spent the evening by the sea, water running between my fingers and toes, the pink light of the midnight sun glowing across the surface like silk. In the far distance I watched a swan gliding between two islands, and always in the sky I could hear the swifts screaming, seemingly oblivious to the time of night, confident only in the knowledge that sunlight meant summer and a time for being loud and alive. I spoke to my fellow swimmer and she was interested to hear about life in Britain. She told me she came here every evening to swim in the summer, to be outdoors and relieve the stresses of daily life. She joked that she couldn't understand why people paid to go in swimming pools.

I swam a few lengths along the line of bright buoys wobbling in the water, all the while watching the little islands, the vanishing sun, the black-headed gulls circling in the sky. It grew darker, and I felt a little sleepy. Leaving behind the sparkling sea of the archipelago, I gathered my things and wandered slowly back to the bus stop, pausing one last time to hear the swifts spiralling through the sky – the loud, glorious harbingers of summer, of growth, of new life.

CHAPTER TEN

Fern Owl

I love cream teas. The crispy topping of a lemon-drizzle cake; thick blobs of fruity jam; a pot of bergamot tea. I'm not bothered about great English traditions like saluting the monarchy or flying flags, but I daren't imagine a world where cream teas are no longer on offer. It's a passion I share with my mum and eldest sister. Over the years we've floated through one tea parlour after another, savouring scones, admiring the decor (fresh flowers are always a hit), and handing over our entire life savings to National Trust properties along the way.

It was a cream tea that first led me to Gilbert White's House in Selborne, home of the English naturalist and reverend who became known for observing and recording the seasonal changes in nature, eventually writing *The Natural History and Antiquities of Selborne*, which has been in continuous print since its publication in 1789. I had a particularly lovely cloth-bound edition I found in a second-hand bookshop in Alton, where I went to college. White was a pioneer in his field, choosing to study living birds and animals in their natural habitat rather than the more common approach of most naturalists at the time, which was to observe dead specimens pinned to a table. Consequently, White was the first person to distinguish the chiffchaff, willow warbler and wood warbler as three separate species based on their different songs. His house in Selborne is now a museum, celebrating his work and the joy of observing nature. It also commemorates the lives of explorer and naturalist Frank Oates, and his nephew Captain Lawrence Oates, a member of Scott's doomed journey to the South Pole in 1911–12, and whose famous last words have echoed down the last century: 'I am just going outside and may be some time.' The expedition did not reach the South Pole before their Norwegian rivals, and most of the group died on the journey home. Oates' body was never found.

Although Selborne is only a few kilometres from my hometown, I don't remember visiting the place until I was in my twenties, when my mum and I decided to explore the house and indulge in a cake-based afternoon. The kitchen garden was in flower and the grounds stretched out under a blue sky, so we expended our energy walking through the 12 hectares of ancient parkland, admiring the ha-ha and the wonderful 2D cut-out of Hercules, a cost-cutting decoration that, from a distance across the meadow, looks like a 3D sculpture. Afterwards we explored the house and were making our way greedily towards the tea room when I spotted something on the mantelpiece in the little parlour. Was it a cuckoo? No: too mottled, with plumage like rotting leaves in February. I looked closer and realised it could only be a nightjar.

Stuffed and mounted, with faded feathers and a glazed expression, this squat taxidermy creature was barely recognisable as the bird that over hundreds of years had carved itself a place in folk history. The scientific name for nightjar is *Caprimulgus europaeus*, the first part translating roughly into 'goat' (*Capri-*) 'sucker' (*-mulgus*). Goatsucker! A vampirical name for a bird that only eats insects, although with their narrow eye slits and ink-black pupils, they can give off a murderous vibe. The name originates from an ancient European belief that nightjars stole the milk from goats' udders, and they were even accused of pecking the hides of cattle and causing a disease called puckeridge, which is actually caused by warble flies laying their eggs under the surface of the cattle's skin. These folk tales became so entrenched that for centuries people still disliked the nightjar, also known as the Flying Toad, Nighthawk, Moth Owl and Fern Owl – the latter frequently used by Gilbert

White in his writing. He reflects ruefully on the power of prejudice and superstition:

> The country people have a notion that the fern owl is very injurious to weanling calves … but the least observation and attention would convince men that these birds neither injure the goatherd nor the grazier, but are perfectly harmless and subsist alone on night insects … It is the hardest thing in the world to shake off superstitious prejudices: they are sucked in as it were with our mother's milk … and make the most lasting impressions, become so interwoven into our very constitutions, that the strongest good sense is required to disengage ourselves from them.

Why has the nightjar been so unfairly disliked for so long? Perhaps due to its squashed, slightly uncomfortable shape at rest, which makes it look like a doorstop. Perhaps due to those dark eyes, large gaping mouth for catching moths, or even the spiky whiskers that line the edge of its beak. Or perhaps it's because the nightjar is rarely seen but always heard. On warm summer nights, an evening walk through an isolated heathland will be peaceful … until suddenly, from the depths of the ferns, will emerge the unmistakable call of the nightjar, a strange, unearthly churr like the vengeful ghost of a Japanese horror film.

In April we found ourselves sailing across the English Channel towards northern France, where our nightjars were making their final stop before heading home for the summer. Nightjars are migratory birds, spending the short summer months on the heathlands, moorlands and open woodland of Britain before leaving at the end of August to overwinter in Africa. In spring they make their way back up through Nigeria, Mali, Morocco, Italy and France, before crossing the Channel and returning to Britain to breed, completing the migratory cycle. There are two species of nightjar in Europe but only one in the UK and, despite (or because of) our historic persecution, I felt protective of our birds. Where did they go when they left our shores? What kind of habitats did they live in when they weren't focused on breeding? We were sailing to one of their holiday chateaux to explore temporary landscapes, ones through which they were only passing on their way back home. Our sailing was an overnighter, and as we sipped apple martinis the sun dipped below the waves, a vast expanse of salt water spread around us on every compass point. Tomorrow we would wake on French waters and discover where our nightjars had been sleeping.

Just after I started secondary school we visited northern France for a few days on a school trip. We stayed in one of those brilliant hostel-type complexes created for students, full of ping-pong tables and Orangina vending machines. The itinerary, designed to supplement our history, geography, French language and cultural skills, took us far and wide across Normandy.

During the trip we visited the Bayeux Tapestry, a 70-metre embroidered linen cloth that had been hanging in France for almost 1,000 years. It depicts the events leading up to the Norman Conquest of England and is perhaps most famous for its image of Harold Godwinson, the last Anglo-Saxon king of England, being killed with an arrow to the eye. Whether or not this is how Harold died has been debated by historians almost since the tapestry's creation, but some believe the idea was deliberately invented by the Normans in an attempt to legitimise William the Conqueror's victory and seizure of the English crown, claiming Harold was 'struck down' by God as punishment for betraying William in the lead-up to the Battle of Hastings.

The tapestry is a magnificent object, a symbol of the complexities of human civilisation, reminding us not only that borders and nations are a man-made concept, but that entire countries thrive on the diversity of their people. Today it seems strange to hear people crying out for an older, more nostalgic version of Britain, and I can't help wondering to which point they would like to return. To the Palaeolithic, when we all migrated over from Africa? The Neolithic, when farming was introduced from the Middle East? To the Iron Age, when the so-called Celts arrived from Poland, Austria and France? How about the Renaissance, when ideas born in Italy fuelled our arts, science, politics and architecture, and the work of pioneering Arab astronomer Al-Battani helped develop the study of trigonometry? John Donne once wrote: 'No man is an island, entire of itself; every man is a piece of the continent, a part of the main.' We have so much to gain from connecting with others, and so much to lose by cutting ourselves away. In the end it is only living things that matter. As an 11-year-old girl, of course, nothing of that kind entered my head. I liked the colours and the amusing expressions on their faces, and I liked using my newly gained Latin skills to try and translate some of the words embroidered through the scenes. Above Harold's sad little head it reads: '*Hic Harold rex interfectus est*', translated as 'Here King Harold has been killed'.

There is another panel in the tapestry, without battle or blood or glorious coronations, which seems to glow with natural wonder, a warm simplicity; a piece of history so vivid you can almost taste the reverence in the Latin inscription above it: '*Isti mirant stella*.' Roughly translated it reads: 'The people look in wonder at the star.'

The star in question is not a star at all, but you can forgive the people of early Britain for thinking so. It is a short-period comet made of dust and volatile ices like water, carbon dioxide and ammonia. More specifically, it is Halley's Comet, the only comet visible to the naked human eye, and the only one we might have the chance to see twice in our lifetimes. Sadly I won't get this chance, as the last appearance came six years before my birth, although the next is due around 2061 when I will be 69 and still heartily outpacing death. One of its earlier recorded appearances came in 12BC, only a few years before the assigned date of the birth of Jesus, and some theologians have suggested this might even explain the idea of the Star of Bethlehem. The original creators of the tapestry played fast and loose with the chronology of this section: although Halley's Comet is shown just after the scene depicting the coronation of King Harold, in reality it appeared about four and a half months later. Historians believe this was to add extra gravitas to Harold's coronation, with the comet representing a sign of divine judgement from God, foreshadowing the evil that would arise after he betrayed William.

Something about this small embroidered comet seems to reach out to us, a fragment of an older life surviving into the modern age, long after we had worked out what comets were and how they are actually relatively unimportant to our daily lives. To these people, however, they might have believed a comet to be a message from God, a spirit sent by the devil, or even a harbinger of the apocalypse. How wonderful that it was stitched into this tapestry by a quiet embroiderer in the corner of a room all those years ago, so that when that comet finally returns to our skies, despite all the technological advances achieved by then, we will witness the same natural brilliance and still share the awe captured by that simple utterance: 'The people look in wonder at the star.'

The next morning we awoke in our cabin, gathered our things and drove the camper van down the ferry ramp and into Saint-Malo, a pretty port city with old granite walls and cobbled roads. From here we travelled to our first destination, Lake Guerlédan, a place I had read about online that was known to be a hiker's paradise. It is in fact a man-made lake, the largest of its kind in Brittany with a surface area of 4km^2, originally constructed to power the nearby Guerlédan dam. Unsure what to expect when visiting northern France in the middle of April, we arrived in the car park and walked over to the cliff edge overlooking the water. That this was a man-made habitat was almost unbelievable. A shimmering palette of blues and greens stretched out before us, the air vibrating with out-of-season quietude, the sky gleaming with swallows just back from Africa. But beyond that, away into the dark forest rising up from the shoreline and encircling the water like a hazelnut husk, a voice floated out and up towards the sky, an echo across the water that seemed to awaken something buried in my earliest memories.

Cu-ckoo.

Two notes, and I was thrown back to my six-year-old self, standing in the patch of ancient woodland that was then my grandparents' garden. Here was where my sisters and I played for years under the dappled canopies of oak and beech leaves, where we made a rope swing and discovered moss-smothered statues in a derelict corner of the garden. It was a vast space, full of wonder and joy and horror, all the emotions a child feels when faced with the magnitude of nature. I remembered a monkey puzzle tree planted as a gift one year; an outbuilding with an old ping-pong table blanched in cobwebs; a sundial standing in a topiary circle that never told the correct time; a dead rabbit covered in flies beneath a rhododendron bush. At one end of the garden we visited the Fairy Hole with Grandad, where we would collect coins left as presents – as long as we always left one behind to prove we weren't greedy. At night we could hear owls and bats in the sky, and Nana would leave a plate of lamb bones out for the fox.

Cu-ckoo.

And there, hiding somewhere in the trees, was the cuckoo. We used to listen to it calling through the forest as we sat in the sunshine, eating hot sausages from the BBQ – the sausages were always ready first so we had something to nibble while everything else was cooking. The cuckoo came in spring, and called all day from some hidden corner, always unseen. Male cuckoos are thought to have absolute pitch, singing always in the key of C; their song starts as a descending minor third when they first arrive, but the interval gets wider as the season goes on, moving from a major third to a fourth, so that by June there is almost no harmony to it at all. But still he sings on, that clockwork voice reverberating through the trees long past the evening and into nightfall.

The common cuckoo has gained an unsavoury reputation in the natural world due to its unique breeding habits. They are known as brood parasites, which means they lay their eggs in the nests of other birds such as dunnocks, reed warblers, pied wagtails and meadow pipits, leaving the unwitting foster parents to raise the cuckoo chick as their own. Adult cuckoos will sometimes mimic sparrowhawks to raise the alarm among smaller birds, distracting them long enough for the female to lay eggs in their nests. It's an amazing adaptation and, like all natural creations that seem slightly unfair, we can't help associating the cuckoo with a little naughtiness. On the darker side, young cuckoo chicks will push host eggs and even live chicks out of the nest to ensure their own survival. It might go against our laws of morality, but it's a highly successful way of reproducing, and a quick Google search will reveal quite amusing photographs of tiny dunnocks feeding adolescent cuckoos almost three or four times their size, bound to care for their monstrous children under the iron grip of maternal instinct.

Sadly, the call of the cuckoo is a much rarer sound in Britain today, and according to research published by the British Trust for Ornithology (BTO), cuckoo numbers in Britain have dropped by 65 per cent since the early 1980s. There doesn't seem to be one exact reason, but like so much of our wildlife, field studies suggest their decline is linked to the health of the whole ecosystem. One theory is that climate-induced shifts in the breeding times of host species have reduced the number of nests available to parasitise; some of these species have brought their breeding forwards by five or six days, which can have both positive and negative effects on the ability for cuckoos to find suitable nests at the right time.

Despite downward trends in Britain, however, the common cuckoo is still classed as 'Least Concern' on the IUCN Red List of potentially threatened species, with a native range of 130 countries and over 100 different host species recorded. In northern France, where the name first originated from the Old French word *cucu*, the cuckoos of Lake Guerlédan seemed to be plentiful, their unmistakable calls following us for our entire 10km hike through the woods and farmlands nearby. It isn't just their nest-stealing habits that class them as vagabonds of the natural world; they are also impossible to find. Everywhere we went, there came the maddening call of the cuckoo, but to discover its hiding place was impossible. Dave said it was like watching a rainbow; so easy to detect from afar, but try to get close enough to touch it and it disappears, always slightly out of reach.

By the end of the day I had gone a little cuckoo-manic, trying and failing to find our cuckoos. Everywhere we went I could hear the bird calling, but no matter how hard I strained to see through the trees and up into the topmost branches, I could never find him. In my delirium I was reminded of the Wise Men of Gotham, an old folk story from medieval England in which a group of villagers feigned imbecility to prevent a visit from the unwelcome King John (who also appears as the stroppy lion from the greatest Disney film ever made, *Robin Hood*). The story goes that King John had intended to travel through the village of Gotham in Nottinghamshire to build a hunting lodge there, but at the time any road the King had travelled on was then made a public highway by law. The people of Gotham did not want a highway through their village, but they were unable to directly disobey the King's wishes, so when the royal messengers arrived they pretended the whole village was mad. A number of absurd tasks were publicly carried out, successfully persuading the messengers to report back to the King that the whole area should be avoided. They tried to drown an eel in a barrel of water, threw cheese rounds down the hill in the hope they might roll their way to market in Nottingham, and built a hedge around an old bush on which a cuckoo was sitting, to prevent it flying away so the summer would never end. On finding itself hedged in, the cuckoo immediately flew off – much to the dismay of the Wise Men – but the Cuckoo Bush Mound still stands in Gotham on top of what is believed to be a Neolithic or Bronze Age tumulus.

By evening we had resigned ourselves to the fact that the cuckoo was a bird to be heard and not seen, and retreated to the camper van for a cup of tea and beans on toast. The heat of the day lingered on as the sky grew darker, so afterwards we hopped down to the shore of the lake and took a dip in the water. It was freezing. If freshwater bodies are shallow enough, they can quickly absorb the warmth of the sun even this early in the year. But the lake was deep and cold. I plunged my head beneath the surface and felt the spring sweat leave my body, the icy water brushing through my hair and into the very pores of my skin. I pushed my head up but kept my shoulders below the water; the surging numbness meant that it felt warmer to stay submerged. I floated in the water, eyes to the cloudless sky that had now turned indigo. At one end of the sky, far from where the sun had set a while before, the stars had started to gleam, specks of white against the fading light. I gazed around at the lake, the forest, the wildflowers and wooden boats. The night air was silent but for the sound that had followed me all day:

the taunting, joyful, clockwork call of a most clever bird, echoing through the trees like a sylvan siren.

Cu-ckoo.

Cu-ckoo.

Cu-ckoo.

The next day we explored the hills and towns of Brittany, consuming vast quantities of crêpes and coffee before driving up to the Monts d'Arrée in Finistère, a small but ancient mountain range reaching its peak at Roc'h Ruz just 385 metres up. This was tonight's sleeping spot, in keeping with our general philosophy of finding secret, beautiful places to park up, moving if asked and leaving no negative trace on the landscape. It was already growing dark by the time we reached our destination, a long, solitary track leading away from the mountain road and down towards a small village. The landscape here could be called bleak, but in truth it was just unforested. The horizon had darkened into a pink shadow fragmented by clusters of grey cloud, and the land was almost barren but for the black, twisted silhouettes of gorse and bracken pressed against the sky. It was a heathland habitat, beloved at dusk by nightjars, bats and birds of prey, but behind us was also a forest, and inside a wren was singing its fiery, rattling warble to the gloaming.

Dave made a start on the food while I went to explore the forest. It was not entirely wild, but more a strange mixture of cultivated land overgrown with wild plants. The floor was carpeted with ferns and bracken, but in front of me stood a line of fruit trees, a seemingly abandoned orchard that was now a feasting den for the birds and mice that lived here. I tackled my way through a bramble thicket and moved around the corner to find a lone deer who was startled and scampered away into the night. Further out beyond the trees, the land returned to sparsity, bracken and gorse, and I felt this was good nightjar territory: the perfect spot for a bird to wriggle down into the ground and make a nest. There were no nightjars calling then, but it was only April; by July these mountains would be ringing with the churring call that mesmerises us all on summer nights. Perhaps they were lying here anyway, alarmed by my presence but staying motionless to avoid detection. Perhaps they were sleeping further down the mountain, under a spiked gorse bush or beneath a frond of green bracken.

Having chased cuckoos through the forests of Lake Guerlédan all day, it was liberating not to have my camera around my neck. I like photography; it's useful to capture beautiful birds or landscapes, but I chiefly use it to illustrate my writing or to brighten up a blog. I'd rather look with my eyes, but I'm too easily seduced by the temptation to see the world through a lens rather than enjoying the moment as it is, there in front of me in 3D. Aside from the flash and my trail camera, I don't have the right equipment to take photos in the dark, so at that moment I felt free to wander through the trees in search of crepuscular life: a strange shuffle in the leaves, the clap of a pigeon disturbed in the canopy, the damp scent of dark earth.

I walked back to the camper van parked snugly under the trees, the orange glow of Dave's culinary activities seeping out through the windows like a lantern in the night. Tonight we would eat pasta and play Bananagrams, and there the cuckoo would be, lingering in the woodland just a few metres down from where we would later fall asleep. Even then he called out to the darkness; an endless chime; a hook to catch hold of my soul and lift me into spring.

On our final night in France we slept by the beach at Normandy, the sand scattered with shells and cuttlefish bones, the odd bottle top and bits of dried bladderwrack infused with salt. A low tide and a low sun; to the west the sky was stained like blood, a marbled canvas of tangerine, crimson, electric purple. The waves washed into the shore and returned to the sea, and at the water's edge a gang of wading birds shuffled through the sediment, jabbing at the ground in search of small things to eat.

I thought we might be standing on one of the D-Day beaches, and imagined the horrors of what had happened below my feet; of who had lived, who had died, and who had fought to rid Europe of a darkness greater than anything found in the natural world. Like many people my age, I don't agree with war as a solution to the world's problems. It belongs in the past; a primitive, reactive step that indulges our basest instincts and disregards all of the intellectual progress we have made as a species. How can conflicts be resolved through violence? It's a strange idea, although I am, of course, clueless to what the alternative is. I'm not an anthropologist or a politician or a sociologist; I have no idea how best to deal with terrorists or poverty or any of the causes for modern warfare today. But I believe that most conflicts are rooted in greed and fear; we are conditioned to expect a high standard of living with lots of material goods, and rather than sharing that with others, we are fed scaremongering tales by the media that encourage us to see other people as 'enemies', when in actual fact they are just trying to survive in the best way they think possible. Sometimes their way of doing things will clash with our own morality, but how can violence be a sustainable way to stop such complex problems? The real answer to conflict has perhaps not revealed itself yet, but in some intricate, long-term way, it must surely be rooted in love and understanding rather than in hate and ignorance.

We stayed awake long into the night, watching the waves roll up the beach and back out to the ocean, out towards England and the south coast where we lived just a few kilometres away. That night we drank bottles of French dry cider effervescing with the aroma of apple blossom and autumn sun, and on the side we ate a slobbering slab of pungent cheese found in a nearby delicatessen. It would be the last piece of cheese we would eat, as we had decided that the next day we would finally commit to something we had thought about for a long time: going vegan.

France seemed an appropriate place to say farewell to dairy, and what a final feast! A creamy, semi-soft hunk that had half-melted in the heat, encased in a washed rind and stinking to high heaven. An absolute dream for a cheese fiend, which I had been ever since working on the Waitrose cheese counter when I was 16. It was cheese, in fact, that had prevented me from committing to veganism for so long. I had turned vegetarian around five years ago for purely environmental reasons, but over the years I learned more and more about the relationship between agriculture, people and the earth. Meat had been easy to give up, and although I loved eggs, I thought they would be easy too. But cheese was my favourite food group – the fouler-smelling, the better. I was still determined to have a taste of *casu marzu*, the traditional Sardinian sheep's cheese that contains real living maggots – a delicacy so far removed from modern health-and-safety standards that it has been banned by the EU and is now only available on the Italian black market.

One evening I read a book about the evolution of the human species and how we first domesticated animals, and it was then that I decided I wanted to finally give up eggs and dairy. The fact that I am alive on this planet, that a certain selection of atoms formed together to give me 100 years of life and joy and love on earth, made me realise that it was impossible for me to take that away from another living thing with a clear conscience, even though domesticated animals were tamed by humans in the first place. I could no longer mourn the loss of wildlife, the catastrophe of climate change, the suffering of other animals, the ecological state of the world, and still cling to bad habits as if they were impossible to change. To know I am reducing my demand on the earth and not exploiting other living things has changed the way I connect with the natural world, and I feel lighter for it, more healthy, more free, more at peace with the world around me.

We dined on cider, artichokes (after seeing them for sale everywhere, we discovered Brittany was famous for them) and our last scoops of soft, French cheese, until the sun disappeared behind the sea and, with the sound of waves lapping against the shore, we fell asleep in the twilight.

My hometown sits just off the border of Hampshire, Sussex and Surrey, and a few kilometres down the road from Iping Common, a beautiful, buzzing heathland filled with bees and flowers and rare, wild things. It was the place I brought Dave during our first summer together, when I was desperate to show him one of my favourite natural events of the year, one that lasts only as long as the nights stay short and warm; by early September it would be finished, and with it the vibrancy of summer.

We drove to the edge of the forest and left the car by the road. It was already 11pm and the sun had set, casting a faint glow across one corner of the sky, blurring into midnight black at the other. Here, empty of clouds, the shadows of space were interrupted by tiny stars and a crescent moon glowing cream in the darkness. We walked from the car into the trees, following a dirt path that wound through the woods in the shade, until after several minutes we emerged into a heathland, the air alive with that arid warmth of bees on heather flowers, rusty bracken crunching underfoot, invertebrates crawling across the dry earth.

There is nothing like a heathland at night, and it never surprises me that Shakespeare chose it as the setting for the Weird Sisters' meeting place in his Scottish tragedy

Macbeth – it's an eerie, evocative place even in daylight. While we are fortunate enough to have 58,000 hectares of heathland in Britain, globally it is rarer than rainforest. In the last two centuries we have lost so much that only 16 per cent remains of the heathland that once grew here; as the Industrial Revolution took hold, factory work became mechanised, cities grew, the countryside was ploughed up, fens drained, trees uprooted, and heathland – also known as the common land – was enclosed. What had once been, quite literally, common land for communities to graze their livestock on and forage firewood, became enclosed in one larger plot of land owned by whoever could afford to buy it. This landlord then controlled how the land was used and, although yields and profits increased, many people felt it signalled the end of the communal countryside. Tenants were no longer able to graze upon the land unless permission was granted, and many poorer people were forced to leave the countryside for factory work in the cities.

Yet, despite the loss, in a strange way we have almost come full circle with our heathlands, as much of the land that was sold off years ago has fallen back into the hands of organisations like the National Trust and Wildlife Trusts, who are nurturing and maintaining them for future generations. Our heathlands are now the product of thousands of years of partnership between humans and their grazing animals, characterised by heather, gorse, fine grasses, wildflowers and lichens. The autumn before, I had met rangers in the New Forest who were removing invasive pine seedlings that, if left alone, would seed out into precious wetland areas and dry them out. The rangers were also cutting down gorse shrubs that had grown so large their value to wildlife was lessened. The pines and gorse were then burned on a controlled fire to destroy the remains. Small sections of the heathland would also be burned later in the season to revitalise the ground and promote new growth.

The rangers worked on a rotation throughout the area to create a mosaic of heathland scrub at various stages of maturity, ensuring a diverse habitat for the abundance of wildlife that found a home in the New Forest. Unlike trees, heathland can't regenerate naturally, so it needed the extra help or it would just revert to species-poor woodland. They used almost exactly the same techniques that had been employed for centuries, and this was done over the winter months to avoid disturbing wildlife like the ground-nesting nightjars. The larger, older shrubs made great perches for birds like the stonechat and Dartford warbler, while young heather flowers attracted the rare silver-studded blue butterfly.

Another part of managing such a diverse habitat is the use of grazing livestock. The New Forest is famous for its curious ponies, greeting visitors and eating windfall apples left out by residents, but you can also find cattle, sheep, donkeys and even pigs roaming through the trees. The variety of livestock means they each graze on different plants in different ways, ensuring the mosaic of habitats continues to flourish. Pigs are released each autumn for 'pannage', a local tradition mentioned in the Domesday Book whereby the pigs are encouraged to eat fallen acorns, chestnuts and beechmasts, which are poisonous to the ponies.

Being such an ancient and intricate fusion of human and natural history, heathlands have been used for centuries as a backdrop for myths and stories. Perhaps the most famous heathland tales of all are those of the Brontë sisters, set against the desolate landscape of the Yorkshire Moors and full of dark twists and doomed romances that many believe capture the hostile beauty of the heath. Authors of classic novels *Wuthering Heights*, *Jane Eyre* and *The Tenant of Wildfell Hall*, Emily, Charlotte and Anne Brontë captured the soul of the moorland in their evocative descriptions of their stories' setting. In *Wuthering Heights*, Catherine Earnshaw lies in her sickbed, but longs to be outside in nature, where she feels she belongs:

> 'Oh, I'm burning! I wish I were out of doors! I wish I were a girl again, half savage and hardy, and free … and laughing at injuries, not maddening under them! Why am I so changed? Why does my blood rush into a hell of tumult at a few words? I'm sure I should be myself were I once among the heather on those hills. Open the window again wide: fasten it open!'

When she is finally laid to rest, she is not buried in a grave, but on a green slope in a corner of the churchyard, 'where the wall is so low that heath and bilberry plants have climbed over it from the moor'.

Iping Common was managed by the Sussex Wildlife Trust, and tonight we had come to listen to the nightjars calling in the dark. I'd been here once before, on a guided walk with the Trust when we listened to nightjars and owls, found bats using echolocation receivers, and learned all about the Bronze Age barrows and burial mounds scattered throughout the reserve. I remembered the route to the centre of the heath, but then we trailed off in search of birds and bats, following a line of trees that ran down one edge of the reserve and into a thick cluster of dark woodland. From here we wound through the trees, got lost, spooked each other and returned to the heathland once more. And there, at last, we caught a sound on the air that stopped us in our tracks, holding our breath as we strained to hear what was emerging from the black night.

It is difficult to describe the call of a nightjar without using the perfectly defined word 'churr', a mechanical, gyrating gurgle that falls between two notes; an eerie call that has perhaps contributed to the nightjar's mysterious reputation. The churring call contains around 1,900 notes per minute, and although it sounds like the bird is calling in two different tones, the ventriloquial manner of its shift from high to low is actually just the nightjar turning his head to project in a different direction. When the churring stops altogether, he is most likely on the move; after a bubbling trill combined with a sharp wing clap, the nightjar has leapt into flight.

With beautiful feather patterns of dead leaves and tree bark, it is almost impossible to find a nightjar in daylight. They are ground-nesting birds, and in order to survive they remain motionless, undetected due to their amazing camouflage against the peaches and browns of heather and bracken. They are so protective of their young that they will even move the entire nest further away if disturbed. A man called Jonnie I once met in the Haslemere Bookshop told me that he had been searching for nightjars with a BTO nest recorder in Surrey, and it was their job to attach a piece of coloured wool to any nests they found so that when the recorder came to write down the location, he would be able to find it more easily. They followed the instructions carefully and secured the wool to the nests, but afterwards the recorder kept complaining that the wool was nowhere near the nests and he'd had to spend ages searching for it nearby. After some investigation, they discovered the nightjar was so outraged at being disturbed by the nest recorders, it had actually rolled each one of its eggs to a new area and rebuilt the nest around them there.

Dave and I stood listening to the bird churring on the air, unsure from which direction it was coming. Like my Finnish woodpecker, the sound seemed to reverberate around the trees and sky so that it was almost impossible to locate the source, especially when we were surrounded by darkness. We moved vaguely towards the direction of the sound, creeping over bracken and moss, desperate not to make too much noise in case the call stopped and that dreaded wing clap signalled his relocation to a new perch. Despite their elusive nature in the daylight, at night they are known to fly close to the heads of those who dare to invade their territory, and I later learned that flashing a piece of white fabric, resembling the white patches on the wings of the male, could be a good way to draw them closer to you.

The nightjar continued to churr from left to right, and we moved closer and closer … until suddenly it stopped. Damn! Unless it happened to fly in our direction it would be difficult to relocate until it started churring again. We waited, breathing slowly and quietly, until a wing clap broke across the moor. As we squinted through the darkness above us, just where the black forest ended and the deep blue of the sky began, a shadow appeared: swift-shaped but large and determined, it rose up in a crescent moon and soared over the trees before disappearing into the darkness once more.

Delighted to have seen my first nightjar in the wild – silhouettes *do* count – we waited once again for the nightjar to begin his call across the hidden landscape of the heath. We didn't have to wait long: somewhere in the darkness to the left, the same sound soon emerged from the trees and floated over on the air, enticing us in like a hypnotist's pendulum.

For a while we continued walking through the heathland, following the sound of the nightjar under the dark sky, bats flying over our heads. Eventually we decided to start heading home, but on the way back we discovered one of my favourite nocturnal species, a tiny creature in the grass beside the path.

Glow-worms are not technically worms, but beetles, with the females glowing greeny-orange on warm summer nights to attract a mate in their grassland habitats. In order to be seen by their male seducers' photosensitive eyes, they clamber up plant stems to cause their light to spread out as far as possible, like a tiny beacon. I've never stumbled upon one in the day – or, if I have, I don't think I'd know it; they are typical beetles, light brown in colour, while the larvae have grey and yellow triangular markings along their bodies, and to find one in daylight would be a small miracle. It's the adult female that gives them their name and reputation, an orb of green suspended in the grass, swaying under the light of the stars.

We stayed and watched the glow-worm for a few minutes, all the while listening to the churr of the nightjar now behind us on the moor. In winter this place would have been frightening – a cold and desolate moorland forest – but in the summer the air was mild, the sky glowing, the wind vibrating with the bird that, for me, had become so closely associated with long days and short nights. We could hear insects moving through the heather flowers, murmurings in the trees and shrubs, the sound of creatures alive in the dark. At last, we left Iping Common and walked back to the car for the short drive home, a warm breeze on our necks and the juddering call of the nightjar following us through the night.

TIF

CHAPTER ELEVEN

The Dart

In the summer of 2017, my friend Viv and I travelled to Rotherfield Peppard in the Chilterns, to take part in a coracle-building workshop with Alistair Phillips, possibly one of the nicest people I've ever met. I had recently developed an obsession with trying out heritage crafts; that year I had also learned to carve a wooden spoon and to forge iron with our blacksmith, so when Viv suggested we build our own coracles, I was more than keen. Also known as a currach, bull boat, quffa or parasil, the coracle is a small boat dating back thousands of years, found in cave paintings from the early Bronze Age but possibly going right back to the Ice Age. I'd seen them before at the farm, where our archaeologist had been building them the traditional way out of waterproof cow hides, animal-hair ropework and locally foraged wood and pitch. But as we actually wanted ours to last and nobody was too bothered about authenticity, we opted for a PVC lining over a woven frame of ash laths. The PVC would protect the coracle from snagging on rocks and ripping a hole in the bottom, which – assuming we managed not to sink – would usually be mendable with duct tape but still a hassle to sort out.

We spent a long day in the sunshine in the balmy warmth of Alistair's garden, constructing our coracles beside chaffinches gobbling up damselflies and red kites soaring over our heads – the house was only 10km from the original holding pens of the kites' successful conservation programme in the 1980s. By the end of the day, we had built our vessels and completed a lesson in using the paddle and sitting correctly so we didn't fall in, and we finished with a float through a lagoon close by, willow leaves trailing in the water, sunlight dancing through the air. It was heaven, and after we said goodbye, Viv drove us back to the farm to store the coracles in one of the tech pods. Since then they have been allowed to stay there in storage, although whenever I go to collect mine it has usually been moved to where it's not in the way. More often than not it's in the roof rafters of the pig palace and I am required to somehow transport a coracle out of a building full of excitable, nibbling piglets.

Adding a coracle to our fleet of transport opened up the landscape to us in a whole new way. Much like having a kayak or a stand-up paddleboard, the coracle has given us access to the hundreds of kilometres of rivers and lakes across the country that are often difficult to access on foot. Powered only by the soft movement of the oar (if you paddle too manically you just go round in circles), its quiet nature has allowed us to get closer to nature and become more immersed in the wild landscapes of our waterways.

In spring we decided to spend a long weekend along the English Riviera, complete with a long-awaited visit to Agatha Christie's house Greenway on the banks of the sparkling Dart estuary, now managed by the National Trust. To finish the weekend, we brought the coracle along in the back of the camper van, hoping to find a quiet stretch of the River Dart for a voyage under the golden light of a summer evening. I love Dartmoor. The last time we visited I bought a book of ghost stories and folk tales that had arisen from the vast moorland over thousands of years of human civilisation. Besides pixies, headless horsemen and numerous appearances by the Devil, Dartmoor had come to be known as the lair of the yeth hound, the spirit of an unbaptised child that had taken the shape of a black, headless dog after death. He was thought to roam the forests at night, wailing into the trees – despite his lack of a head – returning at dawn to his master, the Devil.

Historians believe that *Sherlock Holmes* author, Arthur Conan Doyle, was inspired by the yeth hound when he wrote his bloodcurdling 1902 tale, *The Hound of the*

Baskervilles. This story terrified me from a young age – not because of the story itself, but because we had the 1982 Ladybird version on our family bookcase, complete with a ghoulish green canine apparition on the front. The plot is believed to have been based not only on the yeth hound but also on the legend of a seventeenth-century squire called Richard Cabell who was known as a monstrous man with a thirst for hunting and was rumoured to have murdered his wife. He lived in Brook Hall on the edge of Dartmoor, and on the night he was laid to rest in the sepulchre, a phantom pack of hounds came baying across the moor to howl at his tomb. From that night on, Cabell and his hounds haunted the moors; in an attempt to lay his soul to rest, the villagers placed a huge slab over his tomb and erected a building around it.

However, we didn't only need to look to fiction to work out why Dartmoor had such an eerie reputation. It was also the home of a 200-year-old prison, once thought to be inescapable and originally built to house prisoners of war during the Napoleonic War with France. At this time, thousands of soldiers were taken prisoner and kept in derelict ships called 'hulks', but they were so unsafe that an isolated facility was built on the remote moors of Dartmoor, and at its peak the prison population rose to 6,000 French and American soldiers. More than 1,000 of these died from the horrific conditions inside; their bodies were buried first on the moor, then exhumed and buried again in two cemeteries behind the prison. After the war the prison lay empty for years before being reopened as a minimum-security civilian prison. It became a holding centre for conscientious objectors during the First World War, and was then reopened in 1920 as a facility to hold some of Britain's most violent and dangerous criminal offenders. This gave it the reputation it still holds today for being a merciless, powerful and inescapable fortress – a reputation that is not only inaccurate (prisoners have escaped in the last century) but also disregards the fact that it has an effective rehabilitation and education system where prisoners are able to earn a living, learn new skills and pursue an education to prevent them from reoffending, thereby allowing them a second chance in the outside world.

Dartmoor was undoubtedly an eerie place, but within that eeriness lay its beauty. Dave and I loved visiting the West Country, and I knew that if ever I were to move away from south-east England, I would like to live somewhere between the crumbling, fossilised cliffs of Lyme Regis and the cobbled lanes of St Ives. The last time I'd been down here, however, was not with Dave but to visit my friend Tom the weekend after I'd built my coracle. Tom writes books about cats and people and landscapes, and he had kindly offered to show me the wild beavers that had started breeding on the River Otter, just a few kilometres up the coast in the village of Otterton.

On the River Otter, there are currently around 25–30 Eurasian beavers, a native mammal that disappeared from English waters 400 years ago when they were hunted to extinction under the watch of Henry VIII. In a mysterious twist, the beavers have now returned to the river and nobody knows how, but many people believe they escaped or were deliberately released by conservationists. Now the wild beavers are here, Natural England have issued a licence to the Devon Wildlife Trust for a five-year study to measure their impact in England, and I couldn't wait to get down there and meet the latest additions to our rivers for myself.

Like all wildlife that has evolved in our ecosystems over thousands of years, the beaver has an important role to play in river maintenance in Britain. Contrary to popular belief, they are vegetarian and do not eat fish, but instead snack on plants like the invasive, non-native Himalayan balsam. They also create a wide range of extra habitats within river systems, like small pools and riffles, and by reducing the amount of sediment coming down the river they can help keep their habitats more stable and reduce the risk of flooding. All in all, research has shown that the reintroduction of beavers increases the biodiversity of our landscape, creates and expands precious wetland areas, improves water quality and minimises the impact of high or low water flows.

I was reminded of the gray wolves in Yellowstone National Park in Wyoming, where scientists claim the wolves' reintroduction has restored the ecological balance that disappeared when they were eradicated in 1926. A native resident of the Rocky Mountains, in which Yellowstone is situated, gray wolves were nonetheless seen as a dangerous predator and were hunted without restraint, along with coyotes, bears and mountain lions. After the last individual was shot, the population of elk began to rise and to overgraze on aspen, cottonwood and other woody, deciduous species of vegetation. Park rangers then had no choice but to cull elk numbers to protect the biodiversity of the park – a task that wolves, which had evolved as a natural predator of the elk, would have once undertaken. In the 1990s, a selection of captured wolves were brought in from Jasper National Park in Canada, and left to see if they could tip the balance back to a more natural, sustainable Yellowstone habitat. After some time, conservationists observed that not only had the wolves started preying upon the elk, but their presence drove the elk away from places where they could be easily trapped, such as gorges, valleys and rivers. With fewer elk to overgraze these areas, the vegetation started to regenerate, with the height of some trees increasing fivefold in just six years. With more trees came more songbirds and beavers, and when the beavers built dams along the rivers, otters, waterfowl, reptiles, amphibians and fish arrived, drawn in by the microhabitats created by the beavers. The wolves also killed coyotes, which increased the numbers of their prey species like rabbits and mice, and with their increase, large mammals and birds of prey started to arrive. Even more amazingly, the newly regenerated vegetation helped to stabilise the riverbanks, stop the sides collapsing, and stop soil eroding into the water, changing the entire shape of the river.

If an entire National Park in America could be affected by a handful of wolves, I couldn't wait to see how the beavers might change our landscape to the benefit of wildlife and people. So, after a long ramble around the Dartington estate and a pint of dry cider, we drove to Otterton at dusk and walked along a path leading out of the village and onto the banks of the Otter. Beavers are crepuscular creatures, emerging at dawn and dusk to fulfil their chores before returning to the lodge – marked by a ramshackle heap of twigs – to sleep in the day. We waited for the sun to sink away fully while a woodpecker on the far bank entertained us and then, as darkness fell, we wandered over to the lodge built into the bank and settled down, examining the paw prints and gnawed logs nearby. All was quiet. Another pair of wildlife watchers nearby looked over at us knowingly, their eyes radiating the unspoken question: 'Will they appear tonight?'

Soon the sun had disappeared and the river had grown quite dark … and then *plunk*. Two beavers emerged from the lodge and slipped into the water. If I hadn't known what they were I'd have sworn they were otters, their oily coats gliding through the river like mottled boats. The female of the pair swam to the bank to fetch a strand of balsam, which she crunched off with her teeth and carried back with her to the lodge, head raised above the water like a dog. She then slipped off to an overhanging tree, crawled out of the water and sat roundly on her haunches, pulling leaves from the low branches and gathering them up in her paws.

Meanwhile, the male had disappeared down the river, so we crept along the bank in search of him. He floated along like a log lost in the current. I lost sight fairly quickly but, stepping down closer to the water, soon heard a soft growling sound that forced me to look down. An angry beaver was sitting a metre from my boot, blatantly irritated that I had disturbed his evening routine. In the twilight his fur shone like wet gold, nose wriggling in the air and flat, leathery tail resting on the bank.

For a while we watched them sweep up and down the river, gathering plants and bustling about in the waterway, until it grew so dark that we were forced to leave them behind and return to the village. It was wonderful to see them swimming so freely in the river, a creature that – despite its persecuted history – had found its way back to the British landscape and set up its own pioneering colony like explorers in a new world. A few years ago it was announced that every county in England now had otters in its rivers, a species that had previously declined due to habitat loss and river pollution. The idea that beavers might flourish as well as our otters, might return to our lands and spread across the country, is a warm and happy thought among the many ecological disaster stories we now face.

I'd wanted to visit Agatha Christie's house in Devon for years, so we left early on the Saturday morning after picking up the coracle from the farm (fortunately the piglets were asleep), and drove for three and a half hours down to the Devon coast.

Agatha Christie has been ever-present in my life. When I was younger, Sunday evenings were spent with my mum and sister watching Hercule Poirot or Miss Marple solve crimes in mansions and cottage gardens, surrounded by dazzling costumes, devilish villains and poisoned coffee cups. Nowadays, being a twenty-something freelance creative person, I often rely on Poirot and Marple to get me through my day. Freelancing is heaven in so many ways – and I hope I never have to go back to any other way of living – but sometimes your work looks rubbish, nothing's coming out right and you get stuck in a vortex of frustration. On those days, I put my favourite Agatha Christie audiobook or TV series on and find myself comforted by the same characters and stories that have been with me since I was a child. Suddenly my painting looks prettier, that paragraph sounds better, and I make a cup of tea and carry on with my day feeling strangely reassured by a fictional world of arsenic cocktails and beautiful scenery. It's amazing how much these psychological connections can comfort and revive us – random programmes, books, places, food and people that have hooked themselves into our brains when we were at our happiest, so that we can remember them later and feel the same way.

It isn't just her stories that have inspired me, however, but also the woman herself. The most widely published author of all time, Agatha Christie was fierce, independent and passionate. During the war, she trained to be a nurse and enlisted in the Voluntary Aid Detachment, and acquired a knowledge of poisons during a short time working at the University College Hospital, London. She then threw over a number of suitable young male suitors to shack up with Colonel Archie Christie, a handsome vagabond who could fly planes, ride a motorcycle and dance like a professional. She loved roller-skating, swimming, dancing and travelling, and during a trip to Hawaii she and her husband were among the first Britons to try stand-up surfing. In 1928 she was abandoned by Archie and left to raise her daughter alone, a single mother with an income fuelled predominantly by her writing. Later, during a trip to Baghdad on the Orient Express, she met archaeologist Max Mallowan, 13 years her junior, and married him. They travelled the world together and Agatha often worked on his archaeological digs with him, enjoying a happy marriage until her death in 1976. In the 1971 New Year Honours list, she was promoted to Dame Agatha Christie, just three years after her husband had been knighted for his contributions to archaeology; they remain one of the few married couples where both have been honoured to this rank in their own right. She has now become one of the greatest icons of the British literary canon, and Greenway has been immortalised as the summer home of her and Max's friends and family.

One night in December 1926, shortly after Archie had confessed to having an affair, Agatha Christie climbed the stairs of her home in Berkshire, kissed her daughter goodnight, walked outside to her Morris Cowley and drove away into the darkness. She would not be seen again for 11 days, turning up at a hotel in Harrogate with no recollection of what had happened since her disappearance. The media went berserk, with some claiming it was a publicity stunt for her latest novel and others suggesting her philandering husband had murdered her. Arthur Conan Doyle, a great believer in the otherworldly and paranormal, even took one of her gloves to a clairvoyant to see if the spirits could assist with the search.

At the time, the police decided that, overwhelmed by her husband's disloyalty, she had crashed her car on the steep slope in Guildford (where it had been found abandoned but unscathed) and then boarded a train to the spa town of Harrogate, where she checked into the Swan Hydro hotel under the assumed name of Theresa Neele, her husband's mistress. Years later, with the help of new medical knowledge, a fresh theory has been put forward by her biographer Andrew Norman, who believes she was in the grip of a rare mental condition called a 'fugue state', an out-of-body amnesic trance induced by the stress and trauma of the breakdown of her marriage.

We may never know what happened that night in December 1926, but that doesn't matter. Perhaps the amnesia story was a screen to escape more questions, or maybe she had other intentions that night. Either way, our visit to Greenway was free of the troubles that had haunted her past.

The house was bought by her and Max, who enjoyed a happy marriage and had independently successful careers. Wandering through the house, it felt like the family had just gone out for a walk. It was full of colour and strange, beautiful objects they had collected over the years, as well as trinkets that had inspired various characters and plots within her stories. One embroidered picture entitled 'A Sad Dog' is thought to have been the inspiration for Bob, the Fox Terrier in her novel *Dumb Witness*.

We walked through the gardens at Greenway all afternoon, bursting with foxgloves, fig trees and wild strawberries, a peach house and vinery, hidden pools and statues, and a boathouse overlooking the sparkling Dart estuary. This had been the filming location for the murder in *Dead Man's Folly*, one of my favourite episodes because it features the character Ariadne Oliver, the fictional novelist who is believed to be an autobiographical depiction of Christie herself. Afterwards we sipped elderflower cider in the garden and watched the birds in the beech trees, and then, waving goodbye to the estuary, we finally left the grounds of Greenway and drove inland to Ashburton, the place we had chosen to set sail on our coracle later that night.

As we drove through the forests of Dartmoor, Dave recognised the place as being near an old recording studio where he and his old metal band Vallenbrosa had recorded their first album. As the drummer, there were some afternoons when guitar or vocals were being recorded and he wasn't needed for a few hours, so he would leave the studio and explore the woods outside. One day he managed to get his hands on a scabby dinghy, which he repaired with duct tape so that it would float, and later he found himself sailing down the Dart, vaguely listening to warnings from other people about rapids ahead. He also met a group of kayakers who travelled to the river in the evening to sneak in a night sail down the rapids. They told him the power of the water was so strong during high rainfall levels that it was technically illegal to kayak down certain parts of the river, but that these dangerous parts were the best parts to sail down and were impossible to resist. Fully equipped with head torches and safety gear, the kayakers would head out onto the river under the cover of darkness. Dave didn't venture out in the dark, and thankfully he survived his river adventure long enough to return to the recording studio and finish their album.

We arrived at the edge of a small common and parked the camper van under the trees before unloading the coracle from the back. It's not a large boat but with all that wood it can be heavy to carry, especially after being in the river and laden with the water that had soaked through, and we were usually tired after sailing in it anyway. We changed into our costumes and carried the boat a few metres across a meadow occupied by a flock of jays and then over to the water's edge, together with a towel and a few bottles of cider. The sun had already started setting and was casting a warm glow over the entire place so that every insect buzzing across the surface of the water was illuminated like a firefly.

A few other people nearby were interested in what we were doing, which is one of the best things about it if you're in the mood for conversation. They were finishing a BBQ and about to head home, when one man came over and said excitedly how he had seen a documentary about coracles, and how he had learned they were often built by poorer people who couldn't afford to pay the toll to cross a river; apparently it was cheaper to build a brand-new boat than to pay the extortionate tolls. They chatted to us for a while and then returned to their camp, and we were left in our corner of the riverbank with nothing but the sun, cider and a river to explore. Dave launched the coracle out onto the river and climbed inside, bobbing about for a while as I slipped into the water beside him and went for a swim.

For three hours or more we messed about and watched the sunlight slowly disappear, laughing in the warmth and enjoying the beauty of the river. Dave managed to prop his cider bottle on the coracle seat and he floated along carelessly on the calm current, taking a swig now and then – the ideal mode of transport. The entire bank seemed almost prehistoric, covered in acid-green ferns, flowers, exposed bark and beautiful rock formations, and it reminded me of the exotic plants in *Jurassic Park*.

I swam about contentedly, feeling the silt between my toes and the sun on my back. At one point I stood waiting for him to float back to the bank so I could have a turn at sailing the boat, when I looked down into the water and recoiled in horror. Lying on the riverbed was an eel, grey and slippery amid a shoal of tiny fish and, although it moved slowly across the silt, there was something about it that made me shudder. I'd only recently learned about lampreys. These are ancient, jawless, parasitic fish that evolved 200 million years before the dinosaurs, but they have only recently returned to British waters since industrial-pollution levels have been cleared up. They have a large, round mouth full of teeth, which they use to latch onto animals and suck their blood. Although from a conservation point of view I was happy they were back in our ecosystem, even seeing this little eel was enough to make me skittish. It reminded me of the moray eel in the Jolly Roger Bay level of *Super Mario 64*, which made me scream every time it appeared from the corner of the screen.

This eel was the adult version of the glass eel, which is not a separate species but the term used to describe a young eel, or elver, when it is still in its infancy and the body is almost transparent. As the eel ages, the body grows and solidifies until it looks like this one now did, almost a foot long, silver and butter yellow, with skin so smooth it looked like suede. I wondered how many hundreds of kilometres this one had travelled, how many oceans it had passed through on its way from the Sargasso Sea back to this sunlit section of the Dart, where it now sat quietly on the bottom of the cold river. Eels can grow over a metre long and live for 70 years in the wild and, despite being fish, they are also able to survive out of the water for long periods of time, sometimes crawling through wet grass to reach a new patch of water. We watched it for a while longer until I slipped on a pebble and, disturbed by the cloud of silt, the eel swam away into the deep.

Dave floated back over in the coracle to the far side of the river where he had been watching a grey wagtail hopping around on the rocks, skimming insects off the surface of the water before returning to the rocks to rest. It is an oddly named bird, as its grey back is the least interesting thing about it, giving way instead to the bright yellow belly that flashes like a siren as it glides past. At this time of year, the wagtail would likely be looking after a nest nearby, as they like to nest close to fast-running streams on embankments with plenty of stones and roots where they can find a variety of aquatic invertebrates to sustain themselves, including mayflies, molluscs and beetles.

Attempting to get a closer look, Dave had been using the coracle to its greatest advantage – to creep quietly over without disturbing the wagtail, moving forward with a slow figure-of-eight paddle motion. Unsure if he was a threat or not, the bird remained on the rock, tilting its head from side to side, gazing at the hominoid drifting steadily towards him. At last, with Dave just a metre away on the water, the bird left the rock and disappeared into the evening, leaving the insects for the local Daubenton's bats that are perfectly adapted to skim them off the surface of the water.

The sun had now vanished behind the trees, and we decided we'd spent enough time splashing about on this section of the river. It was time for our twilight journey downstream. The coracle was a small boat but we had discovered it could hold both of us so long as we put a lot of thought into balancing and spreading our weight evenly. The bottom was also deep enough that it could carry the few possessions we had taken with us and, after testing it out back home, we found it difficult to capsize. Full of optimism, we both climbed in, waved farewell to the grassy common and started our voyage downriver.

In a moment the darkening sky disappeared, as the coracle carried us away from the clearing and into a tunnel of twisted trees forming an arch over the river and blocking out the evening light. We passed clouds of flies hovering over the surface of the calm water, and the sound of gentle trickling echoed around the trees so that everything around us seemed full of life and movement, glittering, breathing and flowing.

Even in the coldest of winters, when the surface of the water becomes patchy with ice and frost, it is our rivers that keep moving, the veins and arteries of a landscape that weave through the earth and bring life to stagnant spaces. In spring the rivers thaw and rain falls, and water is carried down from the hills to the sea, bringing fresh oxygen and minerals to the microscopic creatures that sustain our ecosystems. I remember as a child going on pond-dipping sessions in our local nature reserve, sifting through the riverbed to find the dragonfly nymphs and great diving beetles lurking at the bottom, squirming angrily when they were brought to the surface to be held in pudgy infant hands before being released back to the watery depths. Now, here we were in Dartmoor on a midsummer night, revelling in the vitality of the River Dart, carrying ourselves along on it, watching the birds and insects that fed on its nutrients.

We soon observed that the river was not as calm as we thought and so, to avoid an inevitable capsize, I climbed out and swam alongside while Dave carried on paddling. Seconds later, we realised three things. Firstly, the river was far more turbulent than we'd assumed from the tranquillity of the common; after turning a corner and gazing down at the water ahead of us, we spotted the first set of rapids, and the second set after that. Secondly, as it hadn't rained properly in weeks, the river levels were so low that we were floating in just 30cm of water interspersed with massive, slippery rocks; when I tried to half-swim, half-stumble down the river alongside the boat, I was beaten by the current and forced to crawl along on my hands and knees like a feeble sea monster. Thirdly, and most interestingly, we had now sailed so far past the common and into the trees that, with the combined strength of the river and weight of the coracle, there was no way we'd be able to pull the boat out of the river and carry it back. After heaving the coracle onto a lump of earth sticking out of the bank, I waited while Dave climbed out of the river and ran back to the van to dump our valuables, and then watched as he waded back across the water, re-embarked the boat and started the long float down the river while I swam behind.

Within seconds I lost a flip-flop. I slipped on a rock, fell onto my side and watched the green slice of rubber float away in the evening light. Meanwhile, we had reached the start of the first set of rapids and, with nothing to do except make our way down, Dave pushed himself forwards and started swirling down the river in a most melodramatic manner. He hit a rock immediately, which sent him spinning backwards into another one, screaming and laughing in bewilderment, praying he would make it out with the coracle intact. I followed after him in the water, which had become slightly deeper and easier to swim in, but just as we reached the end of the rapids, Dave tipped over and fell into the water while I, scrambling to keep on my feet, scraped my legs across a jagged rock and managed to jump onto another earth ledge in the bank.

I watched as Dave floated away, clinging to the coracle, closely followed by my missing flip-flop, which had become dislodged from a branch by all the commotion. I caught my breath and looked down to see a line of blood dribbling out of a gash on my knee, and as I stood trying to work out how to climb back in and swim down to Dave and the coracle, I put my hand on the branch hanging above me and accidentally shook out five or six white moths. They fluttered down and landed on my leg, which, no matter how many hours of summer sunshine it had been exposed to, still shone as white as in midwinter. I had always embraced my pale skin, but even I was insulted by the idea that six moths had mistaken my leg for the moon.

It isn't a myth that moths are attracted to bright lights like the moon but, although this behaviour has been observed, lepidopterists still aren't sure why. One theory suggests that, as some types of moth are known to migrate, they use the night sky as a navigational tool. Since they would never expect to actually reach the moon while doing this, when they accidentally bump into a lamp or television screen, they become confused and disorientated. Another theory suggests that moths use moonlight as an escape mechanism; if they are disturbed while resting in a bush, their instinct directs them to the moon as this is most likely to be in an upward direction away from danger, rather than down into darkness.

Like bats – and many nocturnal creatures – moths are another species subject to more than their fair share of negative opinion. The fear of moths, known as mottephobia, often comes from the idea that moths nibble holes in our clothes or disturb us by bashing into light fittings until they flutter to the ground in a cloud of dust. Yet moths belong to the same order of insects as butterflies and, of the 2,500 species of moth in the UK, only two can damage clothes. Contrary to popular belief, not all moths are nocturnal, and many of them are as beautifully patterned as butterflies. They are also invaluable pollinators in our ecosystems, helping to keep our honeysuckle, bramble, white campion, thistle and wild-carrot plants thriving.

For these reasons I didn't shake away the moths immediately, but watched them crawl about on my leg before gently poking each one with a finger until they flew away. I looked up, and Dave was waiting for me at the end of the rapids, together with a surprisingly undamaged coracle. I climbed back into the river and stumbled down towards him, and together we continued downstream until we reached another patch of turbulent water.

This time I watched from the shore, where the beech tree roots were pushed deep into the riverbank and I could walk barefoot along the earth. Dave braved the rapids alone, and after another episode of chaotic swirling and crashing into rocks, he finally came to a further quiet stretch of water. I slipped back into the river and paddled over to him, although the water was now deep enough that I could almost swim. Ahead, the final set of rapids was laid out in front of us, a choppy, rock-filled slope that poured down into a beautiful pool with a smooth granite bank that we could use to finally carry the coracle back out and return to land.

The rapids appeared to be brief but treacherous. After a quick pep talk we decided that, while they were too dangerous to swim through, they might possibly be safe enough to sail down together if we paid close attention to our balance. No more laughter – just sensible sailing. We climbed onto a large rock to re-embark, and after several minutes of splashing and sliding around on the algae, we somehow managed to fit both of us in and regain our balance. The water was smooth here, and we focused on the entrance to the rapids, which would only last for 3 or 4 metres before we were safely back in calm waters. At last, exhausted and using all the concentration we had left, we sailed between two huge rocks and headed down into the rapids.

To use an ancient boat like our coracle can be an amazing way to connect with the past, to explore our waterways and feel like a master of the river while you sail along on a vessel that was designed thousands of years ago. It can feel majestic, empowering, sublime, like a Viking marauder or a Celtic warrior, sailing out into the unknown with nothing but your survival instinct for company.

Alternatively, you can sink.

We managed to stay afloat for two seconds. With a loud splitting sound, the coracle snagged on a rock and spun like a tortoise on its back, so that we had barely sailed half a metre before, tipped off balance, we both fell backwards into the water and became fully submerged. Dave scrambled out so that he was sitting back on the riverbed, but after trying to cling to the coracle and pull us both back to safety, he could do nothing but release it and call to me to swim down through the rest of the rapids. The problem was that I was paralysed with laughter and so incapable of moving that, instead of removing myself from the boat, I remained sitting on the bench with no option but to feel the boat sinking lower and lower into the water until it was fully submerged and only my head was poking out. I had become one with the coracle, and we were so heavy that we weren't even moving any more, caught in a nook on the riverbed while the rapids washed past and Dave sat behind me in, quite literally, floods of laughter. We both sat there, useless, breathless and wet, for at least another 20 seconds before I managed to climb out, pull the coracle out of the water and follow it down the rocks into the pool.

We were exhausted. Not only had we traversed three rapids, but my stomach muscles were solid from laughing and it was now almost totally dark. It took every ounce of strength we had left to heave the boat onto the bank and swim after the three empty glass bottles that had fallen out and were now floating quickly down into the next stretch of the river. Amazingly, the coracle was unscathed; the clever lattice structure had acted like suspension as we bashed into every rock available. There was only one victim in our adventure: a towel had fallen out of the boat when we sunk and it was now lost to the depths of the riverbed, where it remains with the eels to this day.

With the coracle pulled onto the bank, we sat by the side of the river, grazed, bleeding, bruised and happy, staring at the stars that had started to emerge from the sky. It was dark now, and we were thankful we'd managed to finish our sailing before the light was entirely lost. I could hear insects hovering around my wet skin, the first of those nocturnal beasts that usually go undetected until their bites come out in itchy red lumps a day later. Dave saw a water vole running along the bank where the trees grew close to the water. We examined my leg, and found the river had washed away most of the blood so that only a thin line trickled down from the wound, diluted by the water droplets that still clung to my skin.

We sat and listened to the roar of the rapids, the gloop of the calmer water lapping against the bank. For us it had been an hour of fun and adventure, but the river – so much bigger than us, and unresponsive to our laughter and bleeding wounds – focused only on moving, flowing on and on until the water reached the sea. It was the lifeblood of the landscape, and we sat on the shore nearby until the droplets cooled on our skin, before wandering back to the camper van with the coracle dangling between our arms like a hammock between two trees.

That night, we fell asleep listening to the last birds singing in the darkness, the sound of the Dart flowing through the forest. With the windows open we could hear nothing but the river – ceaseless, endlessly bringing life, water, micronutrients to the Devon landscape. Nature can be brutal, beautiful or inspiring, but more than anything else it is reliable: always rolling on in the backdrop of our busy lives; always there to immerse ourselves in, to play in or be uplifted by; always there to remind us of our insignificance on earth but also of the joy of being part of something greater than ourselves.

EDWARD THOMAS
POET
TJP

CHAPTER TWELVE

Poet Stone

It hadn't rained for weeks. Outside, the grass had turned ashen and yellow, the only slivers of green being the feathery fronds of yarrow leaves that had somehow stayed hydrated. Every day, the temperature soared into the late twenties and, while it had been glorious to enjoy such a long, hot summer, at night the heat seeped in through the window and I lay awake past midnight, desperate to hear the sweet sound of rain on the glass.

The alarm went off at two o'clock in the morning. It was a Sunday in mid-July, and I climbed out of bed in the dark, wandered into the kitchen, clicked the hob alight and waited for the kettle to whistle. I could hear Dave moving around in bed, trying desperately to wake up after a few short hours of sleep. I made a cafetière of coffee and poured it into a flask, then stumbled back to the bedroom and turned the lamp on to its lowest setting, the gentlest way of establishing we were definitely not going back to sleep. Dave was still horizontal, so I tapped him on the forehead and moved over to the open window to check the outside temperature was still as mild as Port Salut, before pulling on my leggings and a T-shirt, and tapping Dave on the forehead again.

Ten minutes later, we were shutting the door of the flat and tiptoeing out into the night air. There was no breeze but it was deliciously cool and fragrant. I could smell the honeysuckle at the end of the road, although most of the other hedgerow flowers had withered in the heat. With the flask of coffee in my rucksack, we wandered out of our road and into the town centre. The stars were out in all their shining clarity as we moved past the pubs and bars that had locked their doors just hours earlier, past the blacksmith and the silent train station, past Tesco Express and out to the edge of town – towards the dark, looming wilderness of the Hangers.

The Ashford Hangers, to give the place its full name, is an area of woodland to the north of my hometown. The word 'hanger' comes from the Old English 'hangra', meaning a wooded slope, but to the locals this area is also known as Little Switzerland due to its mesmerising views over the South Downs landscape. Most of the trees that grow there are beech, so that to walk beneath the canopy on a summer's day is like walking in water, the sunlight dancing through the leaves and reflecting on the earth below, glancing off your face and shoulders, a sylvan wonderland of light and shadow. It is one of my favourite places on earth to explore, to watch the seasons change and the colours shift, to smell wild garlic on the wind in spring, and in autumn to spot hidden yew berries glowing crimson against their foliage.

At the bottom of one of the slopes, there's an old farm with a charmingly creaky livery where we keep Dave's mum's New Forest pony, Sparkle (although she was recently measured and has now grown so tall she isn't classed as a pony, but a New Forest horse). The livery is home to several horses and their owners, who gather each day to muck out, gossip, hack out into the Hangers and drink coffee in the sunshine. There's an orange cat called Vincent who sleeps on the hay pile, and a leucistic blackbird that's been nesting there for three years in a pile of old tyres behind the school. Leucism is a condition where partial loss of pigmentation causes an animal to have white patches on their skin, feathers or scales, which meant that our blackbird looked like he's been splashed with bleach. Despite this, it had evidently not impeded his ability to survive, and he had become so tame that as soon a wheelbarrow of horse dung was tipped into the manure box, there he'd be, hopping through the pile and picking out the insects for breakfast.

Our livery is a far cry from that demographic of the equestrian world where, despite a shared love of horses, the owners can appear somewhat snooty. These are the people who have given horse-riding a bad name, associating it with wealth and class, as something that only the most affluent members of society can enjoy. While it's undeniable that owning a horse and having lessons is expensive, there are ways to manage it – through sharing, making friends, or simply helping out with the chores. We were so lucky to have found this livery for Sparkle because everyone here was not only kind and helpful, but there was no expectation of what an equestrian should be. As long as everyone cared for their horse's welfare, it didn't matter whether you were a national dressage champion or you simply liked trotting out into the woods once a week. We were all connected by a love of horses and a love for the beautiful countryside around us.

One evening we were riding back from a long hack into the Hangers, through the beech forests, along the chalky ridge and up to Cobbett's View, so named after the English farmer and journalist who loved to explore Hampshire and Surrey on horseback. In his 1830 book *Rural Rides*, he wrote of the place now known as Cobbett's View, a clearing emerging from the woodland that overlooks the rolling Downs from one of the highest points of the Hangers:

> These hangers are woods on the sides of very steep hills. The trees and underwood hang, in some sort, to the ground, instead of standing on it. Hence these places are called Hangers … Out we came, all in a moment, at the very edge of the hanger! And never, in all my life, was I so surprised and so delighted! I pulled up my horse, and sat and looked; and it was like looking from the top of a castle down into the sea.

If we had carried on riding that evening, we could have taken the Hangers Way, a 34km route from Petersfield to Alton that follows the undulating hills and valleys of East Hampshire. We could have taken the horses through the village of Hawkley, home to the Hawkley Inn where we spent many long summer afternoons and cosy winter nights, and along to Selborne where Gilbert White's taxidermy nightjar sat on the parlour mantelpiece. As it was, our horses wouldn't fare well in the dwindling light; Sparkle was bright but frisky, while Hugo, the lovely boy I was riding, was more sturdy but could still be spooked if he lost his nerve. We were heading home before the full velvet night fell upon us, and through the gaps in the beech leaves we could see the sky turning from blue to grey before beginning its descent into faded mauve, when the stars would start emerging like ivory pinpricks.

The air was bursting with the aroma of wild roses, which had tangled themselves into the hedgerow and would later ripen into rose hips, full of nutrients that would keep the birds and mice fed in winter. The world looks different on horseback; we are so used to seeing everything from a metre or two off the ground, but when you are raised up onto a horse a familiar setting is suddenly seen from a new perspective. Instead of your eyes falling on bluebells or cow parsley, the bustling foliage of the hedgerows and forest floor, you are forced to look upwards, into the canopy or out towards the views that are hidden from sight when walking on foot. The South Downs is so diverse in its habitats, so undulating with its high peaks and shadowed valleys, that even looking at it from a couple of metres higher in the air can transform it into a new space.

A similar transformation was happening that night as we walked through Petersfield to the edge of town and over to a green tunnel that led away through a row of houses and into the Hangers. I had never been this way before, but Dave took us through in complete darkness, our steps faintly illuminated by the light of a lately full moon shining through the trees. I was glad to be wearing leggings because the tunnel had been left to grow wild with nettles that brushed against my hands and left a sharp, white lump on my thumb. As we emerged into clearings between the trees, the moonlight revealed the way ahead, so we could see the end of the tunnel before us and the smooth surface of a deserted byroad.

We left the clearing and walked out onto the road, which was lit by a single lamp post glowing feverishly against the black sky. It was only due to the lamp post that we saw the bat. We could hear it rushing past our heads every few minutes in search of insects, but we could only see it when it flashed past the light, a tiny body hurling itself at lightning speed with intricate precision. We watched it for a while, moving up and down the road, and I tested out my rarely used flash-photography skills, managing to capture the bat in motion once or twice before I felt guilty for the disturbance and stopped. On the camera's screen we could see the bat, minuscule and insignificant against the starlit sky, the flash turning it silver, and any elegance it had in motion lost to the unnatural stillness of the photograph.

At last, we left the bat in peace and continued our walk, leaving the boundaries of Petersfield and entering the sleepy beauty of Steep, where we passed the church where my sister got married, and the beautiful, sheep-strewn grounds of Bedales School where my mum worked, described by *Tatler* as 'a bohemian idyll with bite'. We had entered Edward Thomas country, a poet so celebrated in our town that he had his own commemorative Poet Stone placed at the top of the Hangers by the poet Walter de la Mare. It had been there since 1937, when the hill, known as the Shoulder of Mutton, had been dedicated to Thomas 20 years after he was killed in action during the Battle of Arras on the Western Front. It was made of sarsen stone, the same material as the monoliths used to create the mysterious Stonehenge, and the Poet Stone had since become a source of local pride as well as an extremely exhausting hillside to climb.

Born in Lambeth and educated at Oxford's Lincoln College, Edward Thomas worked first as a book reviewer, biographer and literary critic, before moving to Steep in 1906 with his wife and three children. With the encouragement of American poet and friend Robert Frost, he began writing poetry late in 1914, and in the two years that followed he completed around 140 poems before his early death in the First World War. His work had left a lasting mark on Petersfield and the surrounding area, and Thomas returned the affection while he was alive. One of our favourite local pubs, the White Horse, is known to most as 'The Pub with No Name' due to the fact that the wooden signboard is missing, leaving only a metal frame. Few stop to question what happened to it, but the answer can be found in one of Edward Thomas' poems, 'Up in the Wind', where the barmaid relates how the sign used to blow about in the wind, driving her mad until it was supposedly stolen by a thief one night and never replaced:

'Did you ever see
Our signboard?' No. The post and empty frame
I knew. Without them I could not have guessed
The low grey house and its one stack under trees
Was not a hermitage but a public-house.
'But can that empty frame be any use?
Now I should like to see a good white horse
Galloping on one side, being painted on the other.'
'But would you like to hear it swing all night
And all day? All I ever had to thank
The wind for was for blowing the sign down.
Time after time it blew down and I could sleep.
At last they fixed it, and it took a thief
To move it, and we've never had another:
It's lying at the bottom of our pond.'

The poet Ted Hughes called him 'the father of all', and in death Edward Thomas has become one of Britain's finest voices on war and nature, often utilising the cyclical rawness of nature to reflect on the realities of the battlefield, or simply bottling the tranquillity of the South Downs in a handful of words. Many critics believe he expressed a modern recognition of our place as humans within the natural world, and how closely we are connected; how much we rely on nature for our own wellbeing, and how we are not separate or superior to it, but interdependent. For him, the landscape was a constant in a life plagued by depression, with nature providing such a restorative power for him that he pledged to live a pastoral life, here in the foothills of the Hangers, in:

> A house that shall love me as I love it,
> Well hedged, and honoured by a few ash trees
> That linnets, greenfinches, and goldfinches
> Shall often visit and make love in and flit:
> A garden I need never go beyond,
> Broken but neat, whose sunflowers every one
> Are fit to be the sign of the Rising Sun.

It was 'the Rising Sun' that had caused our early-morning ramble through the wilds of Hampshire. Dave and I were both born and raised in Petersfield and had visited the Poet Stone on many different occasions, but neither of us had ever been there to see the sun rise over the Downs. As the weather was so hot and the nights so beautiful, we decided to make use of the July heatwave and walk the 5km to the Shoulder of Mutton in time for dawn, to watch the sun, listen to the birds and be back home again in time for breakfast.

It was Thomas, again, who had given us the idea. Together with his friend Robert Frost, who described Thomas as 'the only brother I ever had', he would regularly walk the Downs long into the night, years before either man made their success as a poet. They would discuss everything a great friendship usually involves, from marriage and war to poetry and wildlife, and it was during one of these night walks that the two men made the greatest decision of their lives. Thomas had been plagued by indecision on whether to join the war effort, despite being anti-nationalist and despising the racism stoked by the press, encouraging everyone to hate German civilians and feed their patriotism with rage. When Frost announced that he would be moving back to America rather than signing up to the war, Thomas had to decide whether to join him there with his family or to join the soldiers in France. The decision process was a long and complex one, but his writing suggests that one particular night walk in the summer of 1914 pushed him further towards fighting in France, when he wrote in his notebook about how he had imagined his countrymen fighting under the same moon as they were now standing:

> A sky of dark rough horizontal masses in N.W. with a ⅓ moon bright and almost orange low down clear of cloud and I thought of men east-ward seeing it at the same moment. It seems foolish to have loved England up to now without knowing it could perhaps be ravaged and I could and perhaps would do nothing to prevent it.

Not only did he feel guilty for not sharing the soldiers' burden, but he also realised that he could no longer remain here in the sanctuary provided for him by nature while the existence of that sanctuary was threatened by the consequences of losing the war. His decision was later cemented by another poem written by Frost, although it was never Frost's intention to send his friend to war. Amused by Thomas' inability to make decisions, he chided him in a poem that is now one of his best known, 'The Road Not Taken', celebrated for both championing the freedom to choose our own path and highlighting the fact that we can never know how our choices affect our lives, as it is not possible to know what could have been. Often when out exploring together, Thomas would dwell on what might have happened if they had gone one way and not the other, sighing over what they might have seen and done. Frost pointed out to his friend: 'No matter which road you take, you'll always sigh, and wish you'd taken another', before sending him this poem two months before he died in battle:

Two roads diverged in a yellow wood,
And sorry I could not travel both
And be one traveler, long I stood
And looked down one as far as I could
To where it bent in the undergrowth;

Then took the other, as just as fair,
And having perhaps the better claim,
Because it was grassy and wanted wear;
Though as for that the passing there
Had worn them really about the same,

And both that morning equally lay
In leaves no step had trodden black.
Oh, I kept the first for another day!
Yet knowing how way leads on to way,
I doubted if I should ever come back.

I shall be telling this with a sigh
Somewhere ages and ages hence:
Two roads diverged in a wood, and I—
I took the one less traveled by,
And that has made all the difference.

When I first learned about the history behind this poem, I couldn't help pitying Thomas, which isn't something

I generally like to do. But if this is a true reflection of the torment he felt with every decision, the regret that consumed him with every choice he made, how much of his life must he have spent thinking of what could have been rather than what was?

As we walked through the Hangers that night, I thought back to autumn and how I had felt so sure that ending my relationship with Dave had been the right thing to do. I was so frustrated, so certain that I was freeing myself from something I didn't want. Now I was back here, walking beside him through the half-light of early morning, and I wondered whether I would change what had happened if I could. Would I tell my past self that I was making a mistake and reassure her that I would be happier staying as we were? No. To quote the most clichéd truth in the dictionary of human discourse, everything happens for a reason; if I had never left Dave, how would I have known what it was like to be apart? Those three months were the most miserable of my life, but if I hadn't experienced them, how would I know how I truly felt? There is no point dwelling on what could have been; I should only focus on what has happened – harvest it, digest it, and use it to continue with my sole purpose on earth: to be happy and free. I refused, however, to only associate Edward Thomas with pity because in every other way he was one of my greatest inspirations, sharing my love for the natural world and acknowledging our place within it. To walk through the same green lanes and hillsides was to step back in time and see the world through his eyes, forgetting there were 100 years between us.

As Dave and I wandered through Steep and finally climbed into the woodlands that lay at the bottom of the Shoulder of Mutton, away from the village and into the wilder realm of the Hangers, we realised it was four o'clock and almost time for the sunrise. With a brisker step we continued our way up towards the hillside, and started to listen out for the other natural wonder we had hoped to experience that morning – one that goes hand in hand with both the rising and setting of the sun. We listened to the stillness in the trees as we walked, the pause between one day and the next, between the past and the undecided future. It was silent, but we knew it wouldn't be for much longer.

The dawn chorus is one of nature's most powerful and captivating spectacles. From four o'clock onwards our gardens, hedgerows and woodlands erupt into a cacophony of song as every male bird attempts to seduce a partner with the greatest opera of their little lives. They time their breeding season to coincide with the warmest part of the year, when food is plentiful and any eggs laid will have a greater chance of survival, but we knew they would also be singing in July, perhaps with less vigour but with just as much heart. This early in the day there was less background noise from the human world, the air was of a different quality and could carry birdsong up to 20 times further. It was still too early for the birds, but our ears were pricked for the first song, waiting for it to emerge from the forest any second. We dipped into a dark patch of woodland where we could hear a soft rustling in the trees, the first stirrings of the morning before the wild world awoke. From here we came to a gap in the trees before emerging onto the side of a deep crevice cut into the land, stretching out from left to right and smoothed over with black tarmac. It was the A3 – a hectic slice of road that wound from London to Portsmouth, historically maintained as a strategically important road to connect the capital city with the main port of the Royal Navy. Despite its heavy use, the majority of the road is only a dual carriageway rather than a motorway, shaped over the centuries to bypass market towns like ours and to create a smooth route for transportation across the south-east.

In recent years, developers have been praised for the construction of the Hindhead Tunnel a few kilometres along from Petersfield, not only reducing bottleneck traffic in the middle of a tiny town, but also diverting traffic underground and away from where the road previously skirted the wildlife-rich woodlands and heathlands of the Devil's Punchbowl. The Punchbowl is a natural amphitheatre and nature reserve overlooking the tumbling Surrey landscape. The origins of its name are numerous, but one story tells how the Devil hurled lumps of earth at the Norse god Thor to annoy him, and the hollow out of which the earth was scooped became the Punchbowl. It has become one of my go-to places to get lost, hiking through the forests and streams hidden in the depths of the Punchbowl itself, and since the tunnel was built, populations of Dartford warblers, woodlarks and nightjars have thrived there. But my favourite corner of the Punchbowl is now the carpet of short grass that stretches in a crescent around the bowl – the small stub of land that is all that remains of the old A3 road. I love to imagine the travellers and traders who had galloped over that road for centuries, the highwaymen who stopped carriages in their tracks, the lovers eloping to faraway places, the soldiers and sailors who passed through on their way to great battles. Now the road had returned to nature, and nature was slowly nurturing it back to life. The edge of the road was bordered by gorse bushes and nettles, and butterflies settled on the short grass to feel the sun on their wings. There were no more cars, no more horses and carts – only the laced boots of hikers, the tyre tracks of cyclists, and the paws of foxes creeping across the reserve at night, listening to the churring call of the nightjar carried on the wind.

We stood over the still-used A3 outside Steep and stared out across the road, which was deserted, empty of the thousands of cars that commuted into London and back every day of the week. We had discovered the one quiet moment when the road was unused, and it felt like we had stumbled into a zombie apocalypse. As much as I find traffic disruptive, it was strange to see such a busy highway so empty, and we moved on without a word, back into the forest and away from the world of tarmac and satnavs. Although the light of dawn had started to creep into the sky, the trees were still so thick that barely anything could penetrate them. The forest floor was bursting with twisted roots and lumps of chalk, some fixed in the ground, others loose and slippery. Unable to see, we were forced to tap into our other senses and, rather than flumping my feet down without a thought, I used them to gauge steady ground, touching them lightly to the floor and only allowing my weight to settle when they told me it was safe to do so.

Onwards we climbed, crossing from the woodland thicket out onto the last stretch of road before the final ascent. From the road we followed a chalk path along the boundary of a field, strands of wheat swaying like grey fibres in the morning breeze, before finally crossing into the last patch of trees that stood at the bottom of the hill. We stopped and gazed up at the path, climbing the Shoulder of Mutton, a hill so steep that I had never mastered it with ease, no matter how fit I might be. I knew I could make it halfway before pausing for a breath and turning to look back at the developing view as we climbed. With the birds still silent in their nests, we started the final hike up to the Poet Stone and the end of our starlit journey.

This was the walk that had inspired Thomas to write his poem 'When First', an ode to his beloved hillside, one that lifted the heart of a man who was, almost certainly, vulnerable to the complexities of life and troubled by what was expected of him. Nature seemed to provide some respite from this, a constant friend, an educator and a source of inspiration. His many walks up to the peak of the hill culminated in the publication of this poem:

When first I came here I had hope,
Hope for I knew not what. Fast beat
My heart at the sight of the tall slope
Of grass and yews, as if my feet

Only by scaling its steps of chalk
Would see something no other hill
Ever disclosed. And now I walk
Down it the last time. Never will
My heart beat so again at sight
Of any hill although as fair
And loftier. For infinite
The change, late unperceived, this year,
The twelfth, suddenly, shows me plain.
Hope now,—not health nor cheerfulness,
Since they can come and go again,
As often one brief hour witnesses,—
Just hope has gone forever. Perhaps
I may love other hills yet more
Than this: the future and the maps
Hide something I was waiting for.
One thing I know, that love with chance
And use and time and necessity
Will grow, and louder the heart's dance
At parting than at meeting be.

I could imagine him here at the base of the path, gazing up at the hillside, desperate to reach the top and look out across the Downs. It would have looked slightly different 100 years ago; now it was bare of trees but full of wildflowers, a flat incline with nothing to obstruct the view, but then it was speckled with towering beech trees that were uprooted in the hurricane of 1987, five years before I was born. I didn't know much about the Great Storm – only that when I was young, we would walk through the ancient woodland around my grandparents' house on Sunday afternoons, and every now and then stumble upon a gigantic oak or beech tree, ripped from the ground and strewn across the forest floor. Most of the time they had become a new part of the ecosystem, a rotting paradise providing nutrients for insects and beetles, a place to hide and for fungi to grow. But sometimes the roots were still intact and a tree would continue to grow aslant, a beautiful, broken anomaly among the rest. One of the trees that had fallen locally was later used at Butser Ancient Farm to carve into a log boat – in the time since the Great Storm it had been left to rest where it fell, and it had seasoned just the right amount to be carved with Bronze Age tools.

A lot has happened since Edward Thomas died on the battlefields of France in 1917, and in many ways the Hampshire landscape has changed almost beyond recognition. Towns and villages have spread to accommodate the sprawl of Londoners looking for a rural retreat from the city, and farming practices saw an overhaul after the war, when farmers were encouraged to maximise yields using artificial pesticides and fertilisers that, although successful in meeting human food demands, also began one of the greatest mass extinctions of wildlife in human history. In 1962, Rachel Carson published *Silent Spring* and revealed to the world just how much damage was being done to the environment and human health by artificial pesticide use. Today, 50 years later, although we are more aware than ever of both the fragility of the ecosystem and our dependence on it, things are not changing quickly enough. In the 2016 State of Nature report published by more than 50 conservation organisations, data collected revealed that 56 per cent of UK species of wildlife are in decline, and around 165 of those species are considered Critically Endangered. The green and pleasant land we are so proud of in Britain is actually a product of one of the least biodiverse countries in the world.

It's easy to become so depressed at these statistics that we give up altogether. Why bother to save the world when it is already so damaged? How can we change the course of the future when almost every powerful industry in the world seems intent on destroying the planet? How can we persuade others to care about nature when we have become so disconnected from it? I'm an optimistic person, and even

I sometimes despair at the situation. But nature is not something we can choose to protect, because the moment we give up on nature is the moment we give up on humanity. The simple fact is that we cannot continue to live on the planet if the environment is damaged. We rely on healthy ecosystems to feed us, clean our air and water, provide oxygen – and these are just the essentials. Can we really live in a world without bumblebees and oak trees, great white sharks and luminescent plankton, capuchin monkeys, polar bears and Venus flytraps? Can we sincerely teach compassion, kindness and responsibility to our children and grandchildren if we ourselves aren't willing to fight for a better world for them?

When I was flying back from India the year before, I watched the documentary *Before the Flood* on the plane (ironic, considering the carbon emissions I was emitting). The film was a collaboration between National Geographic and Leonardo DiCaprio, documenting the devastating impacts of climate change and exploring humanity's ability to reverse what might be the most catastrophic future we have ever faced. I had heard about it many times but, assuming it would make me utterly miserable, I had not been able to bring myself to watch it until I was faced with a 16-hour flight and a limited selection of in-flight entertainment. The film was bleak, shocking and mesmerising, but to my surprise, it was also full of hope. In the final scenes, a climate scientist tells the audience that we can reverse the damage done by global warming and that we can stop the temperature of the earth rising to unsustainable levels. Not only this, but he claims we can actually reduce those temperatures back to where they were, literally restoring the planet to a much more habitable condition for every living creature. There is, of course, an enormous amount of work to be done in order to reach this point, but the important thing is that it is possible. We don't have to surrender ourselves to fear and despair at what the future will inevitably bring because it isn't inevitable. The power to change the world is still up for grabs – and we need to fight for it.

As we ascended the Shoulder of Mutton, past the wildflowers of summer, past the swathes of grass where beech used to grow, I wondered how many of the trees in the forest surrounding the hill had been seen by Thomas and Frost as they walked through the Downs day and night. Most of the great oaks within the woodland would have been here, their thick round trunks swelling over time, grown from saplings sprouted from dropped acorns, which in turn were missed by the red squirrels that lived here before the greys were introduced in the nineteenth century and pushed them out. Just down the road grew the red helleborine orchid, a flower that was now so rare in Britain that it only existed in three places, one being Hawkley Warren in the Hangers. I wondered if it had always been so rare or if, like so many species, its fragility was yet another casualty of modernity. To walk through a place this old was like drifting through a time portal, reminding me of a forgotten England that we could never go back to but bringing hope for a future that did not yet exist.

I was halfway up the hill; as always, being much fitter than me, Dave had gone ahead to the top. From there I had calculated the sun would rise from the western side of the viewpoint, behind the hill and over the top of the trees, so that we would feel the glow of sunrise before we saw the sun itself. I kept walking, one step at a time, and realised the birds were still not singing, despite there being only minutes to go before daybreak. Perhaps the heatwave had hit them hard, too, and they were still asleep – just five minutes more to enjoy the cool stillness of early morning.

Onwards I climbed, breathless and warm, until finally, just metres ahead, I could see the Poet Stone. Engraved on the front was the final sentence from one of Thomas' essays: 'And I rose up, and knew that I was tired, and continued my journey.' And beside the Stone, smiling in the peachy glow of dawn, was Dave, perched on the wooden bench that had been placed there when the Stone was first erected. I stood for a few seconds at the top, clawing for breath in the morning heat before pulling the flask of coffee from the bag and handing it to Dave while I took a photo of the view. The light was perfect – like crushed velvet on the sky, lifting the landscape out of the darkness of night, revealing the forests and rivers, the grass and flowers, the horses, sheep and the shepherd's hut that had become a pleasantly permanent feature of the view. All was quiet and at peace.

And then, out of nothing, deep within the mottled canopy of a silver birch, a voice sung out in the darkness. Ripe, silky notes floated through the air and out into the ether, the first symphony of the morning bursting from the dark feathered chest of a male blackbird. For a few moments the hillside was his stage alone, one solitary voice pouring through the trees. One by one, he was joined by the voices of the other birds, until the air was ablaze with the beautiful chaos of birdsong chiming through the Downs like a thousand silver bells.

It was the sound we had been waiting to hear all morning, and as I turned to Dave, smiling at the riot triggered by one sleepy blackbird, he knelt forwards and asked me to marry him.

I had never felt more at peace, nor more connected with the living, growing world around me, my breath still short from hiking up the hill, my blood warm with oxygen, the light wind of July moving through my hair and across my skin – and the man I loved telling me he wanted to wake up next to me every day for the rest of his life.

We sat together by the Poet Stone for an hour, drinking coffee and watching the sun leap over the beech trees, pouring warmth onto the hillside, onto our bodies and faces, onto the wildflowers that had started to open up their petals for another day of uncompromised light and heat.

We'd each spent our whole lives in the South Downs, this enchanting, half-wild corner of England that had somehow escaped the London sprawl and remained our secret, sacred home, a labyrinth of ancient forests, chalk streams, red kites and rose hips. To the right, I could see the pylon shimmering at the top of Butser Hill, miles away towards Portsmouth, and directly before us lay Steep, the sleeping village that Edward Thomas had called home. To sit here together was like watching all the best moments of our past, two lives lived in the same quiet town, merged together by one chance encounter at a friend's birthday three years before. Life can lead us anywhere and we should never be afraid of where we might go, but there on the hill, with the Downs stretched out before us like a shimmering, emerald lake, I hoped I would never live anywhere else.

With the night over and the warmth of the sun on our backs, we left the Poet Stone behind and walked back down the sloping track towards home – blackbird song on the wind, a cloudless sky above our heads, and the sweet aroma of wild roses drifting through the soft morning air.

Acknowledgements

Thank you …

To Kate Bradbury for helping me shape the idea for this book and Tom Cox for showing me the Otterton beavers. To all who inspired me at the Grant Arms Hotel in March 2018, especially Mark Cocker for his wisdom and guidance, and Louise Gray for a magical morning run with the red squirrels.

To the wonderful team at Bloomsbury who have made becoming an author so utterly enjoyable. To Jim Martin for giving me a chance, to Julie Bailey for believing in me ever since, and to Charlotte Atyeo, Hannah Paget, Hetty Touquet and the rest of the gang for all their hard work.

To all my creative, passionate friends who continue to inspire and educate me, especially Imogen Wood, Victoria Melluish, Hannah Khan, Amber Banaityte, Mark Ranger, Matt Williams, Megan Shersby, Freya Haak, Katy Livesey, Charli Sams, Emily Joáchim, Hannah Rudd, and Catherine and Lizzy Ward Thomas.

To my mum and dad, Ian, Nana, Grandad, Jim and the rest of my lovely family. To Chloë, Hollie and Christie, the ultimate sisters. To my niece Meredith, brother-in-law Simon, and everyone at Butser Ancient Farm, my favourite place in the world. To Chris, Tony, Andy, Jenny, Paul, Eva, Tinks and Sparkle for welcoming me wholeheartedly into the Baker tribe. To our Spanish rescue dog Pablo for making the world even brighter.

And to Dave, my best friend.

Further Reading

Barkham, Patrick. 2014. *Badgerlands*. London: Granta.

Brontë, Emily. 2015. *The Night is Darkening Round Me*. London: Penguin.

Cox, Tom. 2018. *Help the Witch*. London: Unbound.

Geddes, Linda. 2019. *Chasing the Sun*. London: Wellcome Collection.

Masefield, John. 2012. *The Midnight Folk*. London: Egmont.

Mosse, Kate. 2015. *The Taxidermist's Daughter*. London: Orion.

Stoker, Bram. 1983. *Dracula*. Oxford: Oxford University Press.

Wills, Dixe. 2015. *At Night: A Journey Round Britain from Dusk Till Dawn*. Basingstoke: AA.

Yates, Chris. 2014. *Nightwalk*. Glasgow: Collins.

References

Brontë, Emily. 2018. *Wuthering Heights*. Glasgow: HQ.

Byron, George Gordon Lord, and Jerome J. McGann. 2008. *The Major Works*. Oxford: Oxford University Press.

Carson, Rachel. 2000. *Silent Spring*. London: Penguin.

Christie, Agatha. 2017. *An Autobiography*. Glasgow: HarperCollins.

Cobbett, William. 2001. *Rural Rides*. London: Penguin.

Coleridge, Samuel Taylor. 2008. *The Major Works*. Oxford: Oxford University Press.

Conan Doyle, Arthur. 2012. *The Hound of the Baskervilles*. London: Penguin.

Conrad, Joseph. 2007. *The Secret Agent*. London: Penguin.

Du Maurier, Daphne. 2003. *Jamaica Inn*. London: Virago.

Du Maurier, Daphne. 2003. *Rebecca*. London: Virago.

Ekirch, A. Roger. 2006. *At Day's Close: A History of Nighttime*. New York: Norton.

Grahame, Kenneth. 2014. *The Wind in the Willows*. Oxford: Oxford University Press.

Hamsun, Knut. 1998. *Pan*. London: Penguin.

Jansson, Tove. 2003. *The Summer Book*. London: Sort Of.

Leonard, John, and John Milton. 1998. *The Complete Poems*. London: Penguin.

Pullman, Philip. 2011. *His Dark Materials*. London: Everyman.

Shakespeare, William. 2015. *Macbeth*. London: Arden.

Shelley, Mary. 2003. *Frankenstein: Or, the Modern Prometheus*. London: Penguin.

Tennyson, Alfred Lord. 1983. *Idylls of the King*. London: Penguin.

Thomas, Edward. 2004. *Collected Poems*. London: Faber & Faber.

Tolkien, J. R. R. 2007. *The Lord of the Rings*. Glasgow: HarperCollins.

Wells, H. G. 2017. *The War of the Worlds*. Glasgow: Collins.

White, Gilbert. 2013. *The Natural History of Selborne*. Oxford: Oxford University Press.

Wordsworth, William. 2008. *The Major Works*. Oxford: Oxford University Press.

Index